Nico Hotz

**Butane-to-Syngas Processing in Novel Micro-Reactors**

Nico Hotz

# Butane-to-Syngas Processing in Novel Micro-Reactors

## Experimental Investigation of Butane Reforming for Solid Oxide Fuel Cell-based Small-Scale Powerplants

Südwestdeutscher Verlag für Hochschulschriften

**Impressum/Imprint (nur für Deutschland/ only for Germany)**
Bibliografische Information der Deutschen Nationalbibliothek: Die Deutsche Nationalbibliothek
verzeichnet diese Publikation in der Deutschen Nationalbibliografie; detaillierte bibliografische
Daten sind im Internet über http://dnb.d-nb.de abrufbar.
Alle in diesem Buch genannten Marken und Produktnamen unterliegen warenzeichen-, marken-
oder patentrechtlichem Schutz bzw. sind Warenzeichen oder eingetragene Warenzeichen der
jeweiligen Inhaber. Die Wiedergabe von Marken, Produktnamen, Gebrauchsnamen,
Handelsnamen, Warenbezeichnungen u.s.w. in diesem Werk berechtigt auch ohne besondere
Kennzeichnung nicht zu der Annahme, dass solche Namen im Sinne der Warenzeichen- und
Markenschutzgesetzgebung als frei zu betrachten wären und daher von jedermann benutzt
werden dürften.

Verlag: Südwestdeutscher Verlag für Hochschulschriften Aktiengesellschaft & Co. KG
Dudweiler Landstr. 99, 66123 Saarbrücken, Deutschland
Telefon +49 681 37 20 271-1, Telefax +49 681 37 20 271-0, Email: info@svh-verlag.de
Zugl.: Zurich, ETH, Diss., 2008

Herstellung in Deutschland:
Schaltungsdienst Lange o.H.G., Berlin
Books on Demand GmbH, Norderstedt
Reha GmbH, Saarbrücken
Amazon Distribution GmbH, Leipzig
ISBN: 978-3-8381-0701-1

**Imprint (only for USA, GB)**
Bibliographic information published by the Deutsche Nationalbibliothek: The Deutsche
Nationalbibliothek lists this publication in the Deutsche Nationalbibliografie; detailed
bibliographic data are available in the Internet at http://dnb.d-nb.de.
Any brand names and product names mentioned in this book are subject to trademark, brand or
patent protection and are trademarks or registered trademarks of their respective holders. The
use of brand names, product names, common names, trade names, product descriptions etc.
even without a particular marking in this works is in no way to be construed to mean that such
names may be regarded as unrestricted in respect of trademark and brand protection legislation
and could thus be used by anyone.

Publisher:
Südwestdeutscher Verlag für Hochschulschriften Aktiengesellschaft & Co. KG
Dudweiler Landstr. 99, 66123 Saarbrücken, Germany
Phone +49 681 37 20 271-1, Fax +49 681 37 20 271-0, Email: info@svh-verlag.de

Copyright © 2009 by the author and Südwestdeutscher Verlag für Hochschulschriften
Aktiengesellschaft & Co. KG and licensors
All rights reserved. Saarbrücken 2009

Printed in the U.S.A.
Printed in the U.K. by (see last page)
ISBN: 978-3-8381-0701-1

# Acknowledgements

I would like to gratefully acknowledge the scientific, technical, financial, and moral support of numerous people making the realization of the present dissertation possible at the Laboratory of Thermodynamics in Emerging Technologies (LTNT) at the ETH Zurich.

Above all, I wish to express my sincere gratitude to my advisor, Prof. Dr. Dimos Poulikakos, for having given me the opportunity to research in this multidisciplinary project, for his inspiration and guidance, but also for having given me plenty of responsibilty.

I am thankful to Prof. Dr. Wendelin J. Stark for his great guidance, support, and scientific talks as my co-supervisor.

Since the present dissertation contains mainly experimental work, I am thankful to Martin Meuli, Max Hard, and Jovo Vidic for their instant help, whenever hardware problems occurred, fast manufacturing work was necessary, and test rigs crashed. I am likewise grateful to Marianne Ulrich and Sandra Schneider for solving all administrative and financial problems.

I would like to thank my students Christa Jordi, Christian Balmer, Xue Wang, Nurten Koc, and Peter Waffenschmidt for their contributions to my studies and dissertation.

My special thanks go to all my colleagues and friends at LTNT who gave me support, scientifically and personally. Specifically, I like to thank Dr. Dorothea Hollnagel for providing me with the very helpful LATEX template for my thesis.

This work would not have been possible without the motivation and encouragement of my parents Eva and Armin, my brother Manuel and his girlfriend Daniela. I would like to gratefully mention my family including "my" dogs Iowa, Delice, Jade, and Burín.

# Zusammenfassung

In der vorliegenden Doktorarbeit wird die Umwandlung von Butan zu Synthesegas für miniaturisierte Systeme von Festoxidbrennstoffzellen (SOFCs) bei moderaten Temperaturen experimentell untersucht. Diese Arbeit ist ein Teil des sogenannten OneBat-Projekts, das die Entwicklung eines miniaturisierten SOFC-Systems zum Ziel hat und in Kapitel 2 kurz beschrieben wird. Um eine energetisch und exergetisch effiziente Synthesegasproduktion mit hoher Selektivität zu Wasserstoff und Kohlenmonoxid bei verhältnismässig tiefen Temperaturen unter 600°C erreichen zu können, müssen ein sehr wirksamer Katalysator gefunden und innovative Reaktorgeometrien und -strukturen entwickelt werden. Aufgrund der Systemanforderungen wird Partielle Oxidation von Butan mit trockener Luft als primäre Reaktion angestrebt.

In Kapitel 3 wird die katalytische Produktion von wasserstoff- und kohlenmonoxidreichem Synthesegas aus Butan mit Hilfe von Nanopartikeln analysiert, die in einer Flammen-Sprüh-Pyrolyse aus Rhodium/Ceroxid/Zirconiumoxid hergestellt wurden. Der Rhodiumanteil wird von 0 bis 2.0 Gewichtsprozent variiert, und zwei unterschiedliche keramische Fasern (Aluminiumoxid/Siliciumoxid und reines Siliciumoxid) werden zur Fixierung des Reaktormaterials verwendet. Die katalytischen Messungen werden in einem rohrförmigen Festbettreaktor in einem Temperaturbereich von 225 bis 750°C durchgeführt. Das Ziel dieses Kapitels ist es, die Möglichkeit einer effizienten Butanumwandlung für ein SOFC-basiertes Mikrosystem bei Temperaturen von 500 bis 600°C zu zeigen. Die Ergebnisse bestätigen, dass Rh/Ceroxid/Zirconiumoxid ein vielversprechendes katalytisches Material für die Butan-Synthesegas-Umwandlung ist, mit vollständiger Butankonversion und einer Wasserstoffausbeute von 77% bei 600°C. Die katalytische Aktivität des Festbetts hängt erheblich von der Wahl des keramischen Materials ab, Aluminiumoxid/Siliciumoxid oder reines Siliciumoxid, das zur Fixierung des Festbetts verwendet wurde. Dieser überraschende Effekt kann auf eine Wechselwirkung zwischen homogenen und heterogenen chemischen Reaktionen bei höheren Temperaturen im Reaktor zurückgeführt werden.

Um den Brennstoffprozessor kompakter zu gestalten, wurde ein neuartiger scheibenförmiger Festbett-Mikroreaktor entwickelt, der Rhodium/Ceroxid/Zirconiumoxid-Nanopartikel enthält und in Kapitel 4 vorgestellt wird. Dieser Reaktor mit radialer Strömungsrichtung wurde in Bezug auf katalytische Butan-Synthesegas-Umwandlung bei moderaten Temperaturen von 550°C untersucht. Das wesentliche Ziel ist die Entwicklung eines effizienten Butanprozessors, der einfach in ein miniaturisiertes SOFC-System integriert werden kann. Dies wird durch sein kleines Volumen, die einfach gestaltete Geometrie in geschichteten Mikrobauteilen, hohe Kompaktheit, geringen Druckverlust und eine tiefe Reaktionstemperatur erreicht. Rhodium/Ceroxid/Zirconiumoxid besitzt eine exzellente Langzeitstabilität und erreicht sehr hohe Butankonversion und Selektivität zu Synthesegas. Im Vergleich zu einem äquivalenten rohrförmigen Reaktor führt die verbesserte scheibenförmige Reaktorgeometrie zu einer beachtlichen Erhöhung der katalytischen Aktivität bei gleichzeitig 6.5-mal geringerem Druckverlust. Der Anstieg der katalytischen Wirkung wurde detailliert untersucht, indem mögliche Reaktionswege in drei Regionen des radialen Reaktors analysiert wurden. Die Resultate führen zur wichtigen Erkenntnis eines dreifachen Reaktionspfades zur Produktion von Synthesegas auf einem einzigen Katalysator: Die exzellente Selektivität zu Wasserstoff und Kohlenmonoxid für hohe Volumenströme kann durch eine Kombination von Partieller Oxidation, Dampfreformierung und Kohlendioxidreformierung von Butan erklärt werden, was einen direkten und zwei indirekte Reaktionspfade bedeutet.

In Kapitel 5 wird eine neuartige Methode präsentiert, mit der katalytische Nanopartikel mittels eines Sol-Gel-Verfahrens in einen keramischen Schaum eingebunden werden können. Das gesamte Schaummaterial wird in Form einer flüssigen Masse, die den Katalysator enthält, in die vorgesehene Position im Reaktor gebracht, ohne dass ein Substrat mit dem katalytischen Material imprägniert oder beschichtet werden muss. Der so erzeugte Schaum besitzt sehr vorteilhafte Eigenschaften für ein katalytisches Reaktormaterial: einen vertretbaren Druckverlust dank seiner Porosität, hohe thermische und katalytische Stabilität und ausgezeichnetes katalytisches Verhalten. Um die katalytische Aktivität zu untersuchen, wurden Mikroreaktoren mit diesem keramischen Schaum für die Produktion von wasserstoff- und kohlenmonoxidreichem Synthesegas aus Butan verwendet. Der Einfluss von Betriebsparametern wie Reaktorvolumen und der Volumenstrom des Einlassgases auf die Umwandlung des Kohlenwasserstoffes wurde untersucht.

Die Produktion von wasserstoff- und kohlenmonoxidreichem Synthesegas kann auf sehr effiziente Weise durch den Gebrauch von Nanopartikeln mit Rhodiumdotierung selbst bei tiefen Temperaturen von 550°C erreicht werden. Mit Hilfe eines optimierten scheibenförmigen Reaktordesigns und einer vereinfachten Herstellungsmethode mittels eines direkten Sol-Gel-Verfahrens konnte ein Butanprozessor als Teil eines gesamten SOFC-basierten Mikrokraftwerkes für eine tatsächliche industrielle Anwendung realisiert werden.

# Abstract

The present thesis addresses the experimental investigation of butane-to-syngas processing at intermediate temperatures in the context of micro Solid Oxide Fuel Cell systems (micro SOFC systems). This work is part of a project to develop such a micro fuel cell system, the so-called OneBat project, which is described in chapter 2. To achieve energetically and exergetically efficient syngas production with high selectivities towards hydrogen and carbon monoxide at relatively low temperatures below 600°C, a powerful catalyst has to be found and innovative reactor geometries and structures are developed. Due to the system requirements, Partial Oxidation of butane with dry air as oxidant is chosen as the primary reaction.

In chapter 3, the catalytic production of hydrogen- and carbon monoxide-rich syngas from butane by flame-made Rh/ceria/zirconia nanoparticles is investigated for different Rh loadings (0 - 2.0 wt% Rh) and two different ceramic fibers (alumina/silica and silica) as plugging material. The catalytic measurements are achieved in tubular packed bed reactors for a temperature range from 225 to 750°C. The main goal of this chapter is to prove the possibility of efficient processing of butane at temperatures between 500 and 600°C for a micro intermediate-temperature SOFC system. The results show that Rh/ceria/zirconia nanoparticles offer a very promising material for butane-to-syngas conversion with complete butane conversion and a hydrogen yield of 77% at 600°C. The catalytic performance of packed beds strongly depends on the use of either alumina/silica or silica fiber plugs. This astonishing effect can be attributed to the interplay of homogeneous and heterogeneous chemical reactions during the high-temperatures within the reactor.

To improve the compactness of the fuel processor, a novel disk-shaped packed bed microreactor containing Rh/ceria/zirconia nanoparticles is developed and presented in chapter 4. This radial-flow reactor is investigated with respect to catalytic butane-to-syngas processing at moderate temperatures of 550°C. The main goal is the development of an efficient butane processor to be integrated into a micro SOFC system. This can be achieved due to its small size, easily packaged geometry in layered microdevices, high compactness, low pressure drop, and low reaction temperature. It is shown that Rh/ceria/zirconia has an excellent long-term stability and achieves very high butane conversion and syngas selectivity. The improved disk-shaped reactor geometry shows significant advantages in catalytic behavior, at a 6.5 times lower pressure drop compared to an equivalent tubular packed bed reactor. The increased catalytic performance is pursued extensively by investigating possible reaction pathways in three regions of the radial-flow reactor, leading to the significant discovery of a threefold pathway of syngas production on a single catalyst. To this end, it is shown that the excellent selectivities to hydrogen and carbon monoxide for high flow rates are due to the combination of Partial Oxidation, Steam Reforming, and Dry Reforming of butane, indicating one direct and two indirect reaction paths.

In chapter 5, a novel flow-based method is presented to place catalytic nanoparticles into a reactor by sol-gelation of a ceramic foam containing Rh/ceria/zirconia nanoparticles. This method allows for the placement of a liquid foam precursor containing the catalyst into the final reactor geometry without the need of impregnating or coating of a substrate with the catalytic material. The so-generated ceramic foam shows properties highly appropriate for use as catalytic reactor material, e.g. reasonable pressure drop due to its porosity, high thermal and catalytic stability, and excellent catalytic behavior. To investigate the catalytic activity, micro-reactors containing this ceramic foam are employed for the production of hydrogen and carbon monoxide-rich syngas from butane. The effect of operating parameters such as the reactor volume and the inlet flow rate on the hydrocarbon processing are analyzed.

It can be stated that the production of hydrogen- and carbon monoxide-rich syngas is achieved in a very efficient manner by the use of rhodium doped nanoparticles even for a low temperature of $550°C$. By benefiting from an optimized disk-shaped reactor geometry and a more practicable fabrication procedure using a direct sol-gelation method, the potential industrial application of a butane processor can be significantly improved as part of an entire SOFC-based micro-powerplant.

# Contents

Acknowledgements ..................................................... i

Zusammenfassung / Abstract ........................................... iii

1 Introduction ......................................................... 1

2 The OneBat Project .................................................. 5
   2.1 The OneBat System ............................................. 6
   2.2 Current state of the ONEBAT System ............................ 7

3 Syngas production from butane ....................................... 9
   3.1 Abstract ....................................................... 9
   3.2 Introduction ................................................... 9
   3.3 Experiments ................................................... 11
      3.3.1 Catalyst preparation ..................................... 11
      3.3.2 Catalyst characterization ................................ 12
      3.3.3 Test setup for catalyst performance measurements ......... 12
      3.3.4 Testing procedures ....................................... 13
   3.4 Results ....................................................... 14
      3.4.1 Catalyst characterization ................................ 14
      3.4.2 Influence of $Al_2O_3/SiO_2$ and $SiO_2$ fiber plugs on the syngas production . 16
      3.4.3 Influence of the Rh loading on the syngas production ...... 18
      3.4.4 Effect of the reactor tube material ....................... 19

|  |  | 3.4.5 | Temperature profile along the packed bed | 20 |
|---|---|---|---|---|
|  | 3.5 | Discussion | | 21 |
|  | 3.6 | Conclusions | | 24 |
| 4 | **Disk-shaped packed bed micro-reactor** | | | **25** |
|  | 4.1 | Abstract | | 25 |
|  | 4.2 | Introduction | | 26 |
|  | 4.3 | Experiments | | 27 |
|  |  | 4.3.1 | Catalyst preparation | 27 |
|  |  | 4.3.2 | Catalyst characterization | 28 |
|  |  | 4.3.3 | Reactor | 28 |
|  |  | 4.3.4 | Test setup | 29 |
|  |  | 4.3.5 | Testing procedure | 30 |
|  | 4.4 | Results | | 31 |
|  |  | 4.4.1 | Reactor composition and catalyst characterization | 31 |
|  |  | 4.4.2 | Stability test for a disk-shaped reactor | 32 |
|  |  | 4.4.3 | Effect of C/O ratio and total inlet flow rate on catalytic performance | 32 |
|  |  | 4.4.4 | Comparison of a disk-shaped and a tubular reactor | 36 |
|  |  | 4.4.5 | Catalytic reactions taking place in different reactor regions | 37 |
|  |  | 4.4.6 | Effect of catalytic surface area and catalyst site density on catalytic performance | 41 |
|  | 4.5 | Discussion | | 42 |
|  |  | 4.5.1 | Stability test | 42 |
|  |  | 4.5.2 | Effect of C/O ratio and inlet flow rate | 43 |
|  |  | 4.5.3 | Comparison of a disk-shaped and a tubular reactor | 44 |
|  |  | 4.5.4 | Reaction pathways in different reactor regions | 44 |
|  |  | 4.5.5 | Effect of catalytic surface area and catalyst site density | 45 |
|  | 4.6 | Conclusion | | 46 |

## 5 Catalytic ceramic foam    47

- 5.1 Abstract .................................................... 47
- 5.2 Introduction ................................................. 47
- 5.3 Experiments ................................................. 49
  - 5.3.1 Preparation of ceramic foam ............................... 49
  - 5.3.2 Reactor .................................................. 50
  - 5.3.3 Catalyst characterization ................................. 51
  - 5.3.4 Test setup ............................................... 51
  - 5.3.5 Catalytic testing procedure .............................. 51
- 5.4 Results ..................................................... 53
  - 5.4.1 Structural analysis of ceramic foam ...................... 53
  - 5.4.2 Catalytic stability test ................................. 56
  - 5.4.3 Effect of inlet flow rate, reactor volume, and space time . 56
  - 5.4.4 Reactor temperature ..................................... 59
- 5.5 Discussion .................................................. 60
  - 5.5.1 Structural analysis of ceramic foam ...................... 60
  - 5.5.2 Catalytic stability test ................................. 61
  - 5.5.3 Effect of inlet flow rate, reactor volume, and space time . 61
  - 5.5.4 Reactor temperature ..................................... 62
- 5.6 Conclusion .................................................. 63

## 6 Conclusions    65

## Nomenclature    67

## List of Figures    68

## List of Tables    73

## Bibliography    75

# Chapter 1

# Introduction

As the market demand for portable electronic devices has increased dramatically in recent years, their functionality and operating times have coincidentally become limited by current battery technology. This growing gap in power demand versus energy supply from batteries leads to a loss in run time, unless battery alternatives are developed to meet the demands of next generation portable consumer electronics and wireless communication devices. The worldwide power demand of mobile devices has been forecast to increase eight times from the year 2000 to 2010 and market research firms expect that 80 million fuel cell-based power devices will be sold by 2012, only accounting for consumer applications [1].

A very promising technological approach to battery replacement is the development of fuel cells to capitalize the enormous volumetric energy density of liquid chemical fuels, being nearly two orders of magnitude greater than that of state-of-the-art secondary batteries [2], which can only be exceeded by nuclear fuels. Another practical advantage of fuel cell systems over batteries is the instant recharging via replacement or refilling of a fuel cartridge [3].

An evident fuel for all fuel cell systems is hydrogen, since it can be converted very efficiently in fuel cells and has a high energy density per mass. Nevertheless, the low volumetric energy density of hydrogen (Fig. 1.1) disagrees with the limited space available in mobile electronic devices. The storage of hydrogen in gas phase at several hundred bars or in liquefied state is difficult to achieve in small-scale applications with low weight.

An auspicious solution are lower hydrocarbon fuels in liquid phase providing a large energy density both volume- and mass-wise and that can be efficiently converted to a fuel gas mixture usable in fuel cells. In the present work, butane is chosen as fuel due to energy densities of 49.6 MJ/kg and 28.3 MJ/L and its easy storage in liquid phase at about 2.4 bar and room temperature. In a realistic industrial application, butane can be replaced by Liquefied Petroleum Gas (LPG), a widely available hydrocarbon fuel consisting of butane and propane. LPG is a waste product of the petroleum industry and is sold world-wide in small fuel tanks or cartridges for lighters and stoves. Butane is used in this study as a representative hydrocarbon fuel to

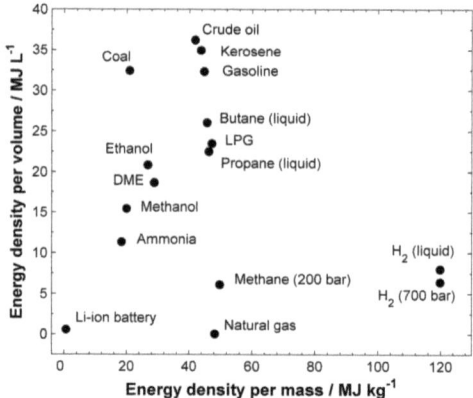

Figure 1.1: Energy density per mass and per volume for different fuels and Li-ion batteries. Sources: Perry's Chemical Engineers' Handbook [4], International Energy Agency (IEA).

simulate the later use of LPG due to its similar chemical and physical properties.

Butane can be used in high-temperature fuel cells such as Solid Oxide Fuel Cells either by direct internal reforming or by external fuel processing. In the first case, the anode of the fuel cell is directly fed with butane and water vapor which reacts via Steam Reforming,

$$C_4H_{10} + 4\,H_2O \rightarrow 9\,H_2 + 4\,CO, \tag{1.1}$$

to a hydrogen-rich gas mixture, where the hydrogen as well as carbon monoxide can be oxidized in the fuel cell and generate electric power. In the latter case, butane is not processed in the fuel cell itself, but in an external reactor. This conversion of butane to a hydrogen- and carbon monoxide-rich syngas is typically achieved by Steam Reforming, Partial Oxidation using dry air,

$$C_4H_{10} + 2\,O_2 \rightarrow 5\,H_2 + 4\,CO, \tag{1.2}$$

or Dry Reforming (also referred to as $CO_2$ reforming),

$$C_4H_{10} + 4\,CO_2 \rightarrow 5\,H_2 + 8\,CO. \tag{1.3}$$

Returning to the idea of small-scale portable power generation, it is obvious that Steam and Dry Reforming are not favorable initial reactions for fuel processing due to their need of water or carbon dioxide input. Unless some kind of exhaust gas recycling separates water or carbon dioxide from the exhaust gas of the fuel cell system and makes them available at the reformer or SOFC inlet, this solution is not practicable for small-scale applications. Additional larger tanks for water (considering the stoichiometry of 1:4 for water) and carbon dioxide (requiring high tank pressures to achieve reasonable volumetric densities) violate the strict volumetric requirements of the entire system already including a butane tank. Therefore, Partial Oxidation is chosen as the preferred initial reaction for this study despite the slightly unfavorable ratio of hydrogen to carbon monoxide compared to Steam Reforming. Since Partial Oxidation is continuously competing with the undesired combustion or Total Oxidation of butane,

$$C_4H_{10} + 6.5\, O_2 \rightarrow 5\, H_2O + 4\, CO_2, \qquad (1.4)$$

Steam and Dry Reforming can occur succeeding Total Oxidation even if dry air is used as inlet gas, as the following chapters will show.

The non-catalytic processing of hydrocarbons typically takes place at temperatures of 1100 - 1300°C in the form of so-called Thermal Partial Oxidation [5]. By using an appropriate catalyst material, the reaction temperature can be reduced by some hundreds of °C, leading to typical temperatures between 700 and 1000°C, e.g. [6–10]. At the beginning of this study, no information was available on catalysts enabling efficient processing of butane at temperatures below 600°C.

The optimal fuel cells to combine with an external butane processor are high-temperature fuel cells, namely Solid Oxide Fuel Cells (SOFCs). Due to its high operating temperature, this type of fuel cell is not poisoned by carbon monoxide, as it happens for low-temperature fuel cells, and SOFCs can even convert carbon monoxide electrochemically. Furthermore, the high temperature typically of 700 - 1000°C [11–13] leads to an increased efficiency of SOFCs compared to low-temperature fuel cells. In recent years, the operating temperature of SOFCs could be decreased by innovative materials as well as modern thin film technologies, leading to so-called intermediate-temperature SOFCs. Currently, 150 mW/cm$^2$ at 550°C [14] can be achieved by research groups within a project to develop a micro SOFC system, the so-called OneBat project (briefly described in chapter 2). It is expected to increase the power density to 200 mW/cm$^2$ already within the next year and finally reach values beyond 350 mW/cm$^2$ by further improving the membranes.

This recent success in the field of SOFC research strongly suggests to develop catalysts and reactor designs for hydrocarbon processors operating with sufficient efficiency at temperatures as low as 550°C. In the present dissertation, this goal is reached by a twofold approach: on the one hand, a highly appropriate catalyst in form of rhodium doped ceria/zirconia is used; on the other hand, this effective catalyst is provided in form of nanoparticles offering an extremely high surface-to-volume ratio and therefore, a very large catalytically active surface in small reactors.

In chapter 3, it is shown that the catalytic reactor is able to produce a reformate gas with 21% mole fraction of hydrogen and 13% of carbon monoxide at reaction temperatures between 500 and 600°C. By using Rh/ceria/zirconia nanoparticles in tubular packed beds, the operating temperature can be dramatically decreased compared to previously existing reactors and catalysts. It is shown that hydrogen- and carbon monoxide-rich syngas is generated by direct Partial Oxidation and by an indirect reaction path consisting of Total Oxidation and subsequent Steam Reforming.

Tubular reactors are not beneficial for integration into SOFC systems. Since the SOFCs consist of a thin, widespread membrane, a disk-shaped geometry for the fuel processor is more convenient. In chapter 4, such a disk-shaped packed bed reactor containing catalytic nanoparticles is presented. The inlet is located in the center of the reactor and the gas flows radially outwards. This increase in cross sectional area along the radial flow direction leads to a significantly decreased pressure drop and provides an important advantage in catalytic performance. As proven in chapter 4, the production of hydrogen and carbon monoxide is not only due to Partial Oxidation of butane, but as well due to Steam and Dry Reforming of butane with water and carbon dioxide previously produced by Total Oxidation. This threefold reaction path on one single catalyst material is highly efficient and underlines the ability of Rh/ceria/zirconia nanoparticles as catalyst for butane-to-syngas processing.

It is clear that loose nanoparticles as catalyst might cause several problems when combined with a micro-machined device produced by clean room techniques, besides possible environmental and health-related issues provoked by nanoparticles. To solve this challenge, a novel method is developed to generate a ceramic foam containing catalytic nanoparticles in a one-step in-situ procedure (chapter 5).

# Chapter 2

# The OneBat Project

State-of-the-art Solid Oxide Fuel Cell (SOFC) systems available on the market are designed for stationary applications in the high power range. Siemens-Westinghouse and Rolls Royce built systems of several 100 kW to the MW region, and systems of HEXIS, Ceramic Fuel Cells Limited, Versa Power, or Topsoe Fuel Cells reach power outputs of 1 to 20 kW. Due to the higher power density of SOFCs compared to other fuel cell types, SOFCs can be considered for powering portable applications with a power range of about 20 to 250 W, such as introduced by Adaptive Materials and Mesoscopic Devices. The technology of these systems is typically based on thick film and bulk processing.

To profit from the high power densities of SOFCs and reduce at the same time the system size, innovative thin film technology and microfabrication techniques can be used to realize the idea of a so-called micro SOFC ($\mu$SOFC) [15–17]. A $\mu$SOFC is expected to be used for applications requiring power in the range of 1 to 20 W as battery replacement in small electronic devices, such as laptops, portable digital assistants, camcorders, medical devices, industrial scanners, or battery charger. The manufacturing of these $\mu$SOFCs is based on thin film technology, microfabrication, and advanced packaging fulfilling complex thermal requirements. So far, the development of such systems is in research status and to date, the published data only relates to the first step of producing of a microfabricated fuel cell based on free-standing thin film SOFC membranes [15–18]. Within the OneBat project, a design for a $\mu$SOFC system is proposed for the first time consisting of a microfabricated fuel cell, a gas processing unit, and the thermal system, as presented previously in [19]. The concept for the system integration is briefly described in the following.

## 2.1 The OneBat System

OneBat is the acronym for a $\mu$SOFC system designed with a base unit of 2.5 W electrical power output and an overall volume smaller than 50 cm$^3$. The hot part of the system is at 350 to 550°C while the outside of the system remains at a safe handling temperature of below 35°C. The system consists of the fuel cell itself, a gas processing unit composed of fuel reformer and exhaust gas post-combustor, a thermal system composed of fuel and air pre-heating unit, heat exchanger, and insulation, and external elements, such as fuel tank, valve, and system control unit. Electrical peaks and start-up are managed by an electrical buffer, e.g. a supercapacitor. Fig. 2.1 shows a schematic illustration of the OneBat system design.

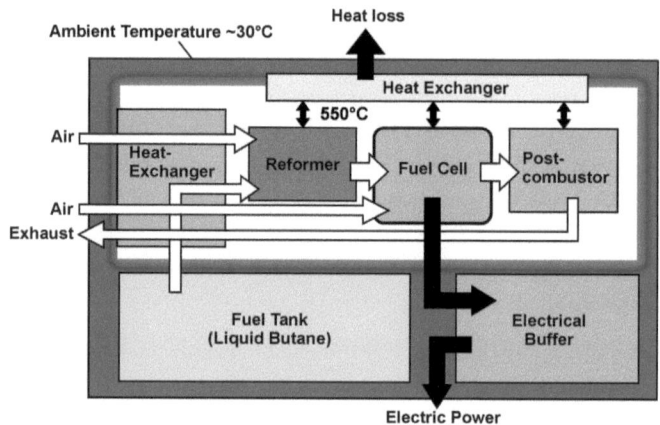

Figure 2.1: Schematic of the OneBat micro SOFC system

In the suggested system design, the fuel cell element is packaged between two substrates, made from Si-single crystal or Foturan®. The gas processing unit is subdivided into the gas preprocessor (reformer) and gas post-processor (post-combustor). Reformer and post-combustor are placed in front and behind the fuel cell element, respectively, and are conceived as modular elements that can be repeated. The insulation encapsulates the micro-system. A micro heat exchanger transfers the heat from the hot exhaust gas to the cold inflowing stream, such that the temperatures of the fluids at the inlet and outlet terminals are maintained as specified.

The modularity of the system enables adaption to different power needs. One single modular element is dimensioned for 2.5 W. By simply repeating these elements, it is possible to address higher power needs for more demanding portable application. The integration of the above mentioned components as well as the micro system fabrication make use of various processes well-established in microtechnology, such as thin film deposition, photolithography, and wafer bonding.

## 2.2 Current state of the ONEBAT System

In order to obtain high cell voltages, dense and pinhole-free electrolyte films are required. This is very difficult with a single layer electrolyte and, in particular, with a vacuum deposition method which usually results in columnar grains in these thin films and, hence, in gas diffusion through pinholes. On the other hand, the double layer electrolyte cell with one thin film made by Pulsed Laser Deposition and one sprayed electrolyte thin film yields in the highest open-circuit voltage of 1.06 V and a power output of 150 mW/cm$^2$ [14].

It is found that each single component of the system, e.g. SOFC, reformer, and post-combustor, can be fabricated and works satisfactory. Fuel cell elements of free-standing membranes of suitable size for a $\mu$SOFC application are fabricated and stacks of these membranes are able to power portable electronic devices. The gas processing unit is able to process the fuel gas to syngas that can directly fuel the SOFC at temperatures around 550°C. Hydrogen, carbon monoxide, and hydrocarbons in the exhaust gas of the SOFC can be converted to water and carbon dioxide by the post-combustor. The large temperature difference of more than 500°C from the hot module to ambient surrounding can be managed according to simulations and thermal insulation experiments. The system design study shows that it is possible to integrate the different subsystems into one complete system which is then fabricated by common microfabrication techniques.

Within the current project, it is shown that a $\mu$SOFC system on-the-chip, such as the OneBat system, is technically feasible. Main results of the different system components prove that a $\mu$SOFC system is an attractive alternative to Li-ion batteries. In particular, it is shown that free-standing, multi-layer thin films of SOFC materials can be processed by thin film deposition and microfabrication. Free-standing ceramic membranes are reinforced by nickel grid structure to reach diameters of up to 5 mm. An open circuit voltage of 1.06 V and a maximum power output of 150 mW/cm$^2$ at 550°C are obtained for single fuel cell elements. Lower operating temperatures and higher power output seam feasible improving the electrolytes and specifically the catalytic activity of the cathodes.

High butane conversion in excess of 90% and hydrogen selectivity higher than 85% are demonstrated in a micro reformer at temperatures of 550°C. Simulations and measurements of the thermal system yield that the aimed temperature difference of about 500°C from inside to ambient temperatures outside is possible for such a small system. Design studies for integrating the different sub-systems into one complete system prove that a $\mu$SOFC system can be produced using conventional power electronic packaging technology. A prototyp will be fabricated as next step.

# Chapter 3

# Syngas production from butane

Parts of this chapter are published in:
N. Hotz, M.J. Stutz, S. Loher, W.J. Stark, D. Poulikakos. Syngas production from butane using a flame-made Rh/Ce$_{0.5}$Zr$_{0.5}$O$_2$ catalyst, APPLIED CATALYSIS B-ENVIRONMENTAL 73 (2007) p. 336 - 344.

## 3.1 Abstract

The capability of flame-made Rh/ceria/zirconia nanoparticles catalyzing the production of H$_2$- and CO-rich syngas from butane was investigated for different Rh loadings (0 - 2.0 wt% Rh) and two different ceramic fibers (Al$_2$O$_3$/SiO$_2$ and SiO$_2$) as plugging material in a packed bed reactor for a temperature range from 225 to 750°C. The main goal of this study was the efficient processing of butane at temperatures between 500 and 600°C for a micro intermediate-temperature SOFC system. The results showed that Rh/Ce$_{0.5}$Zr$_{0.5}$O$_2$ nanoparticles offer a very promising material for butane-to-syngas conversion with complete butane conversion and a hydrogen yield of 77% at 600°C. The catalytic performance of packed beds strongly depended on the use of either Al$_2$O$_3$/SiO$_2$ or SiO$_2$ fiber plugs. This astonishing effect could be attributed to the interplay of homogeneous and heterogeneous chemical reactions.

## 3.2 Introduction

A promising application of fuel cells are miniaturized fuel cell systems generating electric power of the order of a few watts to power small portable electronic devices such as laptops, cameras, and cell phones. Small fuel cell systems using hydrocarbons as a fuel combine the high energetic efficiency of fuel cells with the high availability and easy storage of hydrocarbon fuels [20]. Modern materials for Solid Oxide Fuel Cells (SOFCs) lead to higher efficiencies compared to other types of fuel cells at intermediate operating temperatures, namely in the range of 500 and 600°C [21–23]. An important benefit of SOFCs is the possibility of using hydrogen ($H_2$) and carbon monoxide (CO) simultaneously for electricity production, whereas low temperature fuel cells can not convert CO or even suffer from CO poisoning. This aspect is particularly interesting when fuel cells are supplied with reformate feeds from a hydrocarbon processor which always contain significant amounts of CO. An interesting hydrocarbon fuel for this application is butane, allowing high efficiency for the production of a $H_2$- and CO-rich syngas at moderate temperature similar to the mentioned intermediate temperature of modern SOFCs. Butane can be stored relatively easily in liquid phase at room temperature and low pressure and is widely available.

The major goal of this study was to investigate the feasibility of butane-to-syngas conversion with high $H_2$ and CO mole fractions at relatively low temperatures, while obeying strict limitations of space considering the application of a micro SOFC system. To achieve these demanding requirements, ceria/zirconia nanoparticles with rhodium doping ($Rh/Ce_{0.5}Zr_{0.5}O_2$) made by flame spray synthesis were used as catalysts. Nanoparticles provide open and easily accessible catalytically active surfaces with a high surface-to-volume ratio, as shown by Stark et al. [24]. Zirconia contributes to the thermal stability of the nanoparticles and ceria offers optimal properties for oxygen exchange on the particle surface [25]. The catalytic performance of these nanoparticles was analyzed in a mini packed bed reactor. A desirable reaction path for syngas production from hydrocarbons is Partial Oxidation (POX), written as

$$C_4H_{10} + 2\,O_2 \rightarrow 5\,H_2 + 4\,CO, \quad (3.1)$$

which achieves high yields of $H_2$ and CO. Another effective reaction for hydrocarbon processing is Steam Reforming (SR), where butane reacts with water:

$$C_4H_{10} + 4\,H_2O \rightarrow 9\,H_2 + 4\,CO. \quad (3.2)$$

A second water consuming reaction which might take place in a micro-reactor besides SR is Water Gas Shift (WGS):

$$CO + H_2O \rightarrow CO_2 + H_2. \quad (3.3)$$

A well performing butane processor should show high selectivity towards POX products instead of Total Oxidation (TOX) products of butane:

$$C_4H_{10} + 6.5\,O_2 \rightarrow 5\,H_2O + 4\,CO_2. \quad (3.4)$$

Another objective of this study was the investigation of the effect of different plugging materials of the packed bed on the catalytic performance. Ceramic fibers made of silicon oxide and aluminum oxide ($Al_2O_3/SiO_2$), which is widely used as the sealing material and the substrate of monoliths or foams in hydrocarbon-to-hydrogen reactors, e.g. [6, 10, 26–28], and pure $SiO_2$ were used to form plugs.

An important aspect of catalytic hydrocarbon processors is the combination of homogeneous and heterogeneous reactions simultaneously taking place. Previous studies [29–31] have suggested that oxidative dehydrogenation of propane is a combination of gas-phase homogeneous and catalyzed heterogeneous reactions, where Xu and Lundsford [31] based this interpretation on experiments with $SiO_2$ chips comparable to our $SiO_2$-sand filled tubes. Huff and Schmidt [8] showed that oxidative dehydrogenation of butane in short-time reactors was highly affected by an increase of reactor surface in the absence of noble metal catalysts. Lemonidou and Stambouli [32] confirmed this effect of the surface-to-volume ratio on oxidative dehydrogenation of butane in a non-catalytic $SiO_2$ reactor. The triggering of gas-phase homogeneous reactions by radicals formed by surface reactions during catalytic POX of methane was reported by Campbell et al. [33] and of propane by Silberova et al. [34] and Aartun et al. [35]. The same finding for POX of butane was reported by Marengo et al. [36]. The latter suggested that fast ignition of exothermic homogeneous reactions by the surface of inert packing material such as $SiO_2$ was possible in absence of noble metal catalysts during POX of butane.

The conversion of hydrocarbons to syngas by POX is usually accompanied by TOX followed by SR and WGS, which produce $H_2$ and CO as well. This indirect reaction mechanism was already suggested by Kunimori et al. for butane on platinum [37]. More recently, this autothermal reforming path was investigated by Wang and Gorte for hydrocarbons on Pd/ceria [38]. These studies suggested ceria-supported precious metals as well-suitable catalysts for hydrocarbon reforming catalyzing POX, SR, and WGS simultaneously.

## 3.3 Experiments

### 3.3.1 Catalyst preparation

Ceria/zirconia nanoparticles with optional rhodium doping were prepared in a one-step process by flame spray synthesis described by Madler et al. [39] and Stark et al. [40]. For the ceria/zirconia precursor, cerium(III) 2-ethylhexanoate (12 wt% Ce, Shepherd Chemical Company) and zirconium(IV) 2-ethylhexanoate (18 wt% Zr, Borchers GmbH) were mixed to result in a metal molar ratio Ce/Zr of 1:1 and diluted with xylene to a total metal concentration of 0.8 mol $L^{-1}$. Rhodium(III) 2-ethylhexanoate (UMICORE AG & Co.) was optionally added to the Ce/Zr-precursor such that the calculated rhodium content in a ternary system Rh/ceria/zirconia (Rh/Ce$_{0.5}$Zr$_{0.5}$O$_2$) was 2.0, 0.5, 0.25, and 0.1 wt%. The mixtures were fed (5 mL min$^{-1}$) through a capillary (i.d. 0.4 mm) using a gear-ring pump (HNP Microsysteme) and were dispersed by oxygen at the tip of the capillary (5 L min$^{-1}$, constant pressure drop at the nozzle 1.5 bar, PanGas, tech) following ignition by a methane (1.13 L min$^{-1}$, PanGas, tech)/oxygen (2.41 L min$^{-1}$, PanGas, tech) supporting flame. The burning spray of the flame synthesis reactor was stabilized by a concentric oxygen sheath flow (230 L h$^{-1}$, PanGas, tech). All gas flows were controlled by calibrated mass flow controllers (Brooks). The particles formed in the flame were separated from the off-gas with a glass fiber filter (Whatman GF/A, 25.7 cm in diameter) placed above the flame by the aid of a vacuum pump (Busch Seco SV 1040 C). The thermal stability was tested by sintering as prepared particles at 1000°C for 16 h, under air, using a heating rate of 10°C min$^{-1}$ and measuring the remaining specific surface area.

### 3.3.2 Catalyst characterization

The specific surface area (SSA) of the powders was measured by nitrogen adsorption at 77 K on a Micromeritics Tristar using the BET method (error ± 3%). The phase composition and formation of ceria/zirconia mixed oxides was investigated by X-ray powder diffraction on a Stoe STADI-P2 (Ge monochromator, Cu K$_{\alpha 1}$, PSD detector). To determine the rhodium content, powders were digested in hydrochloric acid (37 wt%, J.T. Baker) and analyzed by flame atomic absorption spectrometry (AAS) on a Varian SpectrAA 220FS (slit width 0.2 nm, lamp current 5.0 mA) applying a N$_2$O (11.0 L min$^{-1}$, PanGas)/acetylene (6.23 L min$^{-1}$, PanGas) flame and measuring absorption at a wavelength of 369.2 nm [41]. Transmission electron micrographs were recorded on a CM30 ST (Philips, LaB$_6$ cathode, operated at 300 kV, point resolution 2 Å). Prior to analysis, the particles were dispersed in ethanol and deposited onto a carbon foil supported on a copper grid.

### 3.3.3 Test setup for catalyst performance measurements

Expanded butane from a liquid tank (PanGas, 3.5, 99.95%) at 2.5 bar was mixed with compressed synthetic air (79% N$_2$, 21% O$_2$, PanGas, 5.6, purity of both species: 99.9999%) from a gas tank (Fig. 3.1). Both flow rates were controlled by Low Delta-P flow meters (Bronckhorst). The flow meters allowed control of the inlet pressure to reach the ambient pressure at the system outlet, such as to be able to operate the reactor slightly above ambient pressure. The butane/air mixture was fed into a packed bed reactor inside an Inconel tube (Alloy 600, i.d. 2 mm) and heated by a tube furnace (MTF 12/38/250, Carbolite). The tube was heated over a length of 30 cm, ensuring isothermal conditions along the packed bed reactor (total length: 11 mm) placed in the middle of the furnace. The assumption of isothermal conditions in the reactor was confirmed by temperature measurements in the packed bed, shown in the appendix.

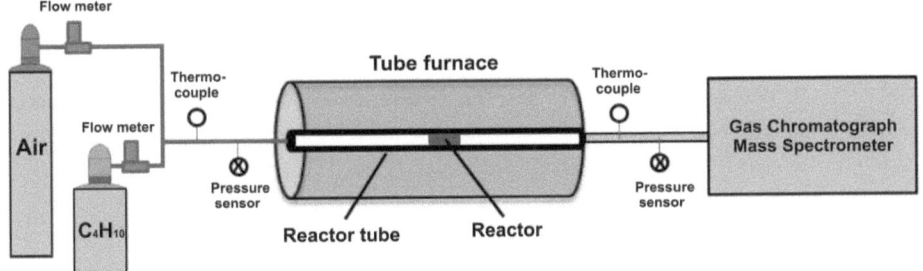

Figure 3.1: Schematic of the test setup for catalyst performance measurements of a packed bed reactor

The product gas leaving the furnace through an Inconel tube was maintained at around 115°C to avoid condensation of $H_2O$. The gas composition was analyzed by a gas chromatograph (6890 GC) coupled with a mass spectrometer (5975 MS, Agilent), using a HP-MOLSIV and a HP-PlotQ column (Agilent), respectively. Helium (PanGas, 5.6, 99.9996%) was added as an internal standard for GC calibration. Under typical run conditions, the molar product gas balances of C, H, and O were closed within 5%.

The packed bed consisted of 22.5 mg purified and calcined $SiO_2$ sand (Riedel-deHaën, average diameter: 0.2 mm) and 7.5 mg $Rh/Ce_{0.5}Zr_{0.5}O_2$ nanoparticles (0-2.0 wt% Rh), prepared as shown before and homogenously distributed onto the sand particles. For reference measurements without nanoparticles, packed beds with 30.0 mg $SiO_2$ were prepared. These porous packed beds were fixed in the Inconel tube between ceramic fiber plugs, as shown in Fig. 3.1. In order to investigate the reactivity of the fibers, two different ceramic materials were used: Ceramic fiber paper made of $SiO_2$ and $Al_2O_3$ (1:1, CT 1260 P, Contherm, mean fiber diameter: approximately 18 $\mu$m, mean fiber length: 5 mm) or pure $SiO_2$ fibers (Riedel-deHaën, mean fiber diameter: 16 $\mu$m, mean fiber length: 5 mm). The catalytic effect of the Inconel tube on the reactions was negligible in presence of $Rh/Ce_{0.5}Zr_{0.5}O_2$ catalyst, as proven by a comparison with a packed bed in a quartz tube (see subsection 3.4.4).

### 3.3.4 Testing procedures

The activity of $Rh/Ce_{0.5}Zr_{0.5}O_2$ catalysts was tested at constant air and butane inlet flow rates of 15.0 and 1.2 mL min$^{-1}$, respectively, leading to mole fractions of $X_{C_4H_{10}} = 7.4\%$, $X_{O_2} = 18.5\%$, and $X_{N_2} = 74.1\%$. This corresponded to a butane mass flow of 0.2 g h$^{-1}$. The C/O ratio or equivalence ratio $\phi$ based on POX was defined as

$$\phi = \frac{2 \cdot \dot{n}_{C_4H_{10},in}}{\dot{n}_{O_2,in}} \tag{3.5}$$

and was kept constant at $\phi = 0.8$ for all measurements, as suggested by previous numerical studies on micro hydrocarbon processors [42, 43]. The entire reactor was heated from 225°C to 750°C at a heating rate of 2.5°C min$^{-1}$ under air/butane flow. The molar composition of the outlet gas was determined every 75°C. The Gas Space Velocity ($GSV$) was defined as the ratio of total volumetric flow rate at the reactor inlet and reactor volume,

$$GSV = \frac{\dot{V}_{\text{gas,in}}}{V_{\text{reactor}}} = \frac{1}{t_{\text{space}}}, \qquad (3.6)$$

which is the reciprocal of the space time $t_{\text{space}}$ [44]. The mean gas velocity at the reactor inlet varied from 0.14 to 0.29 m s$^{-1}$, depending on the reactor temperature, leading to $GSV$ between 15.7 and 32.3 s$^{-1}$ and $t_{\text{space}}$ in the range of 63.7 and 31.0 ms. Based on the results of the GC/MS, the butane conversion $\eta$ of the micro-reactor was determined as the molar ratio between converted butane and inlet butane:

$$\eta = \frac{\dot{n}_{C_4H_{10},\text{in}} - \dot{n}_{C_4H_{10},\text{out}}}{\dot{n}_{C_4H_{10},\text{in}}}. \qquad (3.7)$$

The hydrogen yield was defined as the molar ratio of H$_2$ in the outlet gas and the maximal possible amount of H$_2$ formed at full conversion according to Eq. 3.1:

$$\Psi = \frac{\dot{n}_{H_2,\text{out}}}{5 \cdot \dot{n}_{C_4H_{10},\text{in}}}. \qquad (3.8)$$

The selectivities for H$_2$ and CO were defined as:

$$S_{H_2} = \frac{\dot{n}_{H_2,\text{out}}}{\dot{n}_{H_2,\text{out}} + \dot{n}_{H_2O,\text{out}}} \qquad (3.9)$$

and

$$S_{CO} = \frac{\dot{n}_{CO,\text{out}}}{\dot{n}_{CO,\text{out}} + \dot{n}_{CO_2,\text{out}}}. \qquad (3.10)$$

## 3.4 Results

### 3.4.1 Catalyst characterization

Feeding mixtures of rhodium, cerium, and zirconium carboxylate precursors into a flame spray synthesis unit allowed for continuous production of ceria/zirconia nanoparticles with optional rhodium doping (nominal content 0.5 wt% Rh) at a nominal production rate of 43 g/h. Transmission electron microscope (TEM) images of as-prepared ceria/zirconia $Ce_{0.5}Zr_{0.5}O_2$ (Fig. 3.2 (a)) and 0.5 wt% $Rh/Ce_{0.5}Zr_{0.5}O_2$ (Fig. 3.2 (b)) showed highly regular, crystalline, sharp-edged nanoparticles of 10-40 nm. Addition of 0.5 wt% Rh had no significant influence on the morphology. No metallic rhodium particles could be detected in electron micrographs (TEM) or by energy dispersive X-ray analysis (not shown) indicating a high dispersion of rhodium on the support material. This stays in line with previous studies on Pd/alumina [45], Pt/alumina [46], and Pt/ceria/zirconia [47] synthesized by flame spray pyrolysis that also reported high noble metal dispersions. Both as-prepared materials exhibited a high BET specific surface area (SSA) of 103 $m^2$ $g^{-1}$ for $Ce_{0.5}Zr_{0.5}O_2$ and 106 $m^2$ $g^{-1}$ for 0.5 wt% $Rh/Ce_{0.5}Zr_{0.5}O_2$. The Rh content of the as-prepared samples was measured by flame atomic absorption spectrometry (AAS). Ceria/zirconia contained less than 0.002 wt% Rh (error of measurement) and 0.51 wt% Rh for the doped material, confirming that no Rh was lost during synthesis.

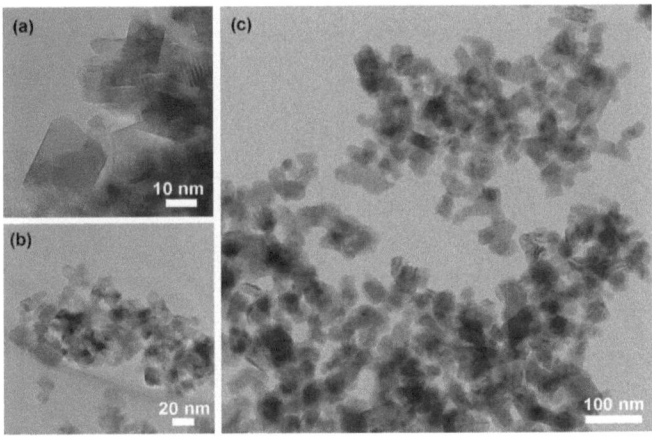

Figure 3.2: Electron micrographs of flame-made $Ce_{0.5}Zr_{0.5}O_2$ (a) and 0.5 wt% $Rh/Ce_{0.5}Zr_{0.5}O_2$ (b) showed regularly shaped nanoparticles with similar morphology. After sintering at 1000°C for 16 h under air, the $Rh/Ce_{0.5}Zr_{0.5}O_2$ (c) displayed marginal particle growth. Sintering necks between adjacent particles stabilized the open, well accessible material.

Severe sintering conditions (1000°C for 16 h [47]) resulted in some sintering but corroborated the high thermal stability of flame-made $CeO_2/ZrO_2$ (Fig. 3.2 (c)), which corresponds well with results of Stark et al. [40]. Sintered samples showed specific surface areas of 28 $m^2$ $g^{-1}$ ($Ce_{0.5}Zr_{0.5}O_2$) and 22 $m^2$ $g^{-1}$ (0.5 wt% $Rh/Ce_{0.5}Zr_{0.5}O_2$). In contrast to studies on Pt/alumina [46] and Pt/ceria/zirconia [47], rhodium had a small influence on the sintering of the support material.

The phase composition of as-prepared or sintered samples was measured by X-ray diffraction (XRD) (Fig. 3.3). Both powders showed broad signals after preparation which could be attributed to a solid solution of cubic metastable κ-$Ce_{0.5}Zr_{0.5}O_2$ [48] as observed in previous studies on flame-made ceria/zirconia nanoparticles [40, 47, 49]. The Rh containing $Ce_{0.5}Zr_{0.5}O_2$ exhibited no additional reflections of the metallic phase, which stays in agreement to the electron micrographs (Fig. 3.2). After sintering of the nanoparticles, signals for cubic $CeO_2$ [50] evolved next to the solid solution and were attributed to partial phase separation [40].

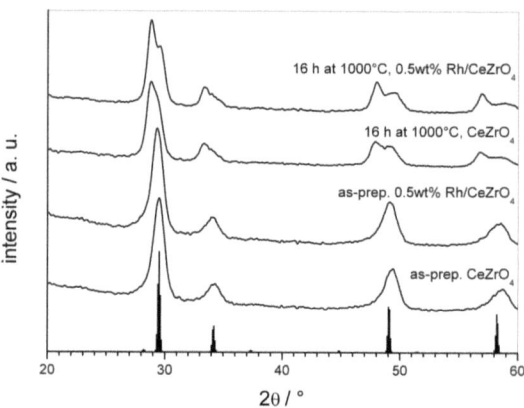

Figure 3.3: X-ray diffraction patterns of as-prepared (bottom) and sintered samples (top). After preparation, the material consisted of a homogenous ceria/zirconia solid solution. The severe sintering conditions resulted in partial phase segregation. Bottom: reference pattern of κ-$Ce_{0.5}Zr_{0.5}O_2$ [48].

## 3.4.2 Influence of $Al_2O_3/SiO_2$ and $SiO_2$ fiber plugs on the syngas production

In Fig. 3.4, the syngas production of packed bed reactors containing $Al_2O_3/SiO_2$ fiber plugs is compared to that of packed bed reactors containing $SiO_2$ plugs. For packed bed reactors consisting of $SiO_2$ sand and ceramic fiber plugs without any $Rh/Ce_{0.5}Zr_{0.5}O_2$ catalyst, the conversion of butane started above 400°C (Fig. 3.4(a)). Between 500 and 700°C, the butane conversion was significantly higher for pure $SiO_2$ fibers than for $Al_2O_3/SiO_2$ fibers. Without catalyst, the amount of produced CO was very low and a small amount of $H_2$ was only produced at 750°C. For reactors containing $Ce_{0.5}Zr_{0.5}O_2$ nanoparticles with 0.5 wt% Rh, conversion of butane started below 300°C (Fig. 3.4(a)). Between 300 and 400°C, the butane conversion was slightly higher when $Al_2O_3/SiO_2$ fiber plugs were used. However, $Al_2O_3/SiO_2$ fibers led to stagnation in butane conversion (at around 50%) up to 525°C, compared to a significantly higher conversion for pure $SiO_2$ fibers with up to 99% at 750°C. The $H_2$ production grew with increasing the reactor temperature up to 375°C, stagnated between 375 and 500°C for $Al_2O_3/SiO_2$ fibers, and finally increased again for temperatures up to 750°C, as it can be seen from the hydrogen yield (Fig. 3.4(a)) and the $H_2$ selectivity (Fig. 3.4(b)). For $SiO_2$ fibers, no stagnation occurred at intermediate temperatures and the $H_2$ production exceeded evidently those of $Al_2O_3/SiO_2$. The CO production started above 300°C and for $SiO_2$, it rose rapidly to a CO selectivity $S_{CO}$ of 73% at 675°C and 79% at 750°C (Fig. 3.4(b)). For $Al_2O_3/SiO_2$ fibers, the CO production increased slower with temperature, reaching high CO selectivities only above 675°C.

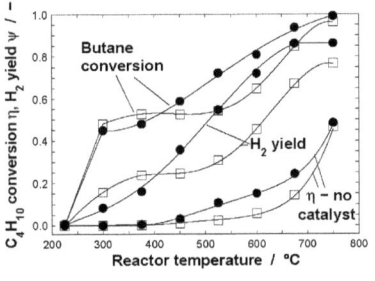

(a) Butane conversion $\eta$ and hydrogen yield $\Psi$

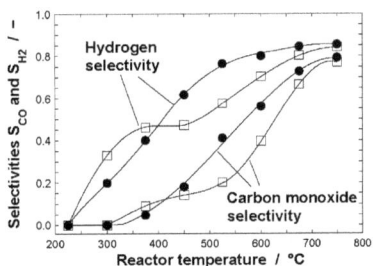

(b) Carbon monoxide and hydrogen selectivities $S_{CO}$ and $S_{H_2}$

Figure 3.4: Butane conversion $\eta$ and hydrogen yield $\Psi$ (a), carbon monoxide selectivity $S_{CO}$ and hydrogen selectivity $S_{H_2}$ (b) as functions of reactor temperature for packed bed reactors with a catalyst loading of 0.5 wt% Rh and for both types of plugs, with $SiO_2$ (•) and $Al_2O_3/SiO_2$ (□) fibers, respectively, for an inlet C/O ratio $\phi = 0.8$. The butane conversion is as well shown for a packed bed without catalyst consisting only of $SiO_2$ sand and both fiber materials. The solid lines through the data points are curve fits.

There were no significant amounts of lower hydrocarbons in the product gas detected. In presence of $Rh/Ce_{0.5}Zr_{0.5}O_2$ catalysts, the mole fractions of lower hydrocarbons were negligible (always below 0.05%), except for methane, which was below 0.7% for $Al_2O_3/SiO_2$ fibers and below 0.3% for $SiO_2$ fibers. For the measurement without any catalytical nanoparticles, the ethane and propane mole fractions were below 0.05% and the ethylene and propylene mole fractions below 0.1%. The methane mole fraction was below 0.9% for this case. Therefore, it can be concluded that in the absence of an $Rh/Ce_{0.5}Zr_{0.5}O_2$ catalyst lower hydrocarbons were formed, but their amount was still insignificant.

The molar product gas balance of C closed within 1% for all measurements, indicating that no significant carbon deposition took place inside the reactor. The occurrence of carbon deposition was also tested by optical inspection. The amount of 0.1 mg of carbon mixed with packed beds (22.5 mg $SiO_2$ sand and 7.5 mg nanoparticles with different Rh loadings) can be optically inspected. For all examined packed beds, no change of color after one measurement cycle (around 400 min) could be detected. Measurement of the weight of the packed bed before and after runs did not indicate the deposition of significant amounts of carbon (detection limit: 0.1 mg, corresponding to 3.3 mg C per g packed bed material).

For both packed beds containing catalyst, complete conversion of $O_2$ could be detected for temperatures above 300°C ($O_2$ mole fractions below 0.05%). At 225°C, no $O_2$ was consumed. Without catalyst, the $O_2$ conversion was below 10% for temperatures below 500°C. Above 500°C, the conversion of $O_2$ increased, reaching complete conversion at 750°C.

### 3.4.3 Influence of the Rh loading on the syngas production

The conversion of butane in packed beds consisting of $SiO_2$ sand, catalyst nanoparticles with Rh loadings from 0 to 2.0 wt%, and $SiO_2$ fiber plugs started below 300°C. For 0 wt% Rh, the start of butane conversion was delayed compared to packed beds containing Rh. The variation of catalyst loading had no significant influence on the butane conversion $\eta$ up to 375°C (Fig. 3.5(a)). Between 400 and 600°C, higher Rh loadings led to remarkably higher butane conversion. For the highest Rh loading of 2.0 wt%, $\eta$ increased rapidly from 56% at 375°C to 92% at 525°C, reaching complete conversion of butane above 600°C. The difference in conversion for materials with 0.5 and 0.25 wt% Rh was rather small. The addition of 0.1 wt% Rh to the ceria/zirconia had a significant effect on the butane conversion. For 0 wt% Rh, the butane conversion stayed between 41 and 53% from 375 to 750°C, whereas 0.1 wt% Rh already led to 86% butane conversion at 750°C.

For all investigated materials containing Rh, the start of butane conversion coincided with the start of $H_2$ production below 300°C. For 2.0 wt% Rh, the hydrogen yield $\Psi$ increased almost linearly to 86% at 675°C, staying at this maximum value for higher temperatures (Fig. 3.5(b)), and the $H_2$ selectivity $S_{H_2}$ reached saturation at 86% above 525°C (Fig. 3.5(d)). At temperatures below 400°C, the Rh loadings of 0.5 and 0.25 wt% showed nearly the same hydrogen yields. On the other hand, for higher temperatures a catalyst with 0.5 wt% Rh allowed a fast rise in $H_2$ production. Above 600°C, the addition of more noble metal did not significantly improve the $H_2$ yield. For 0 wt% Rh, no $H_2$ was produced at any investigated temperature.

In presence of Rh, the production of CO started later than the conversion of butane and the production of $H_2$, namely above 350°C. Above 400°C, the CO production rose rapidly with large differences in CO selectivity $S_{CO}$ between 500 and 650°C for different Rh loadings (Fig. 3.5(c)). The increase of Rh loading from 0.25 to 0.5 wt% had only a small effect on the CO production. The CO production converged to a maximum of $S_{CO}$ of 83% for all catalyst loadings at 750°C. Without Rh, a small amount of CO was produced at temperatures above 375°C leading to a selectivity $S_{CO}$ between 9 and 17%.

As in the case of the previous subsection, no significant amounts of lower hydrocarbons were detected in the off-gases. The mole fractions of lower hydrocarbons were always below 0.05%, except for methane, which was below 0.9%. The C mass balance was closed within 1% and no

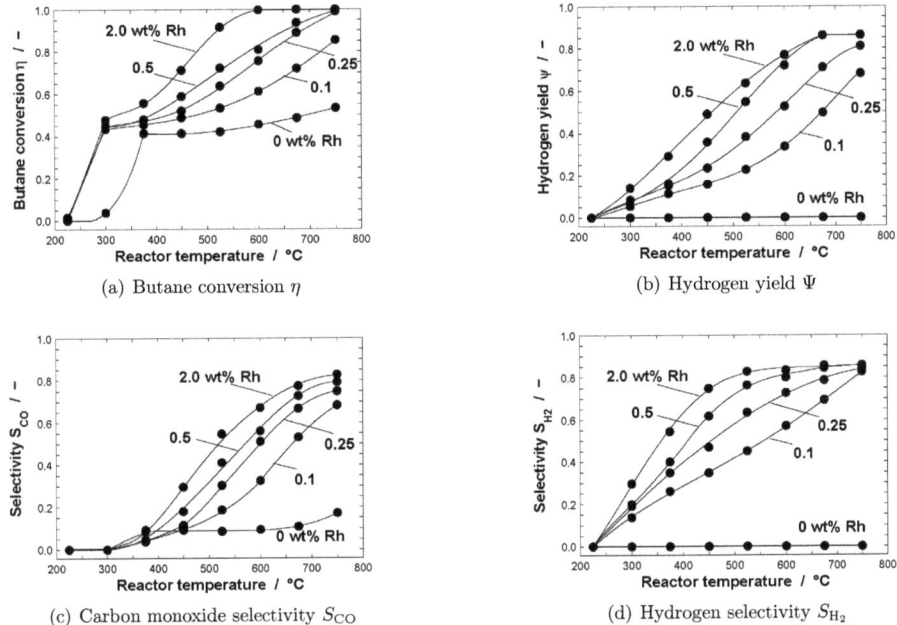

Figure 3.5: The influence of Rh loading on $Ce_{0.5}Zr_{0.5}O_2$ on butane conversion $\eta$ (a), hydrogen yield $\Psi$ (b), carbon monoxide selectivity $S_{CO}$ (c), and hydrogen selectivity $S_{H_2}$ (d) for an inlet C/O ratio $\phi = 0.8$, using plugs of $SiO_2$ fibers. The solid lines through the data points are curve fits.

carbon deposition could be detected optically, indicating the presence of less than 3.3 mg C per g packed bed material. In addition, measuring the weight of the packed bed did not show any carbon deposition either (detection limit: 0.1 mg).

For reactors with Rh loading, the $O_2$ conversion ranged from 99 to 100% for temperatures above 300°C. At 225°C, no $O_2$ was consumed. For temperatures above 500°C, the $O_2$ mole fraction was always below 0.05% with practically complete $O_2$ conversion, and for high Rh loading, no $O_2$ could be detected in the off-gas. Without any Rh loading, hardly any $O_2$ was converted below 300°C. Above this temperature, the $O_2$ conversion reached 95 to 99%.

### 3.4.4 Effect of the reactor tube material

All results presented before were conducted for packed beds in an Inconel (Alloy 600) tube due to its easy handling and high robustness. To prove its feasibility as a reactor material in the sense of chemical inertness, reference measurements with the packed beds inside quartz tubes were performed. These results showed no significant effect of the Inconel on the production of syngas. In Fig. 3.6, characteristic results for two packed beds containing catalytic nanoparticles with 2.0 wt% Rh and $SiO_2$ fiber plugs inside an Inconel and a quartz tube are compared. Except for a small difference in butane conversion and $H_2$ production at 300°C, both reactors showed almost perfect agreement. Therefore, it can be stated that the Inconel tube does not effect the syngas production by $Rh/Ce_{0.5}Zr_{0.5}O_2$ nanoparticles in a significant way for temperatures above 350°C.

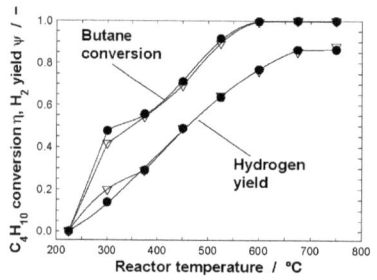
(a) Butane conversion $\eta$ and hydrogen yield $\Psi$

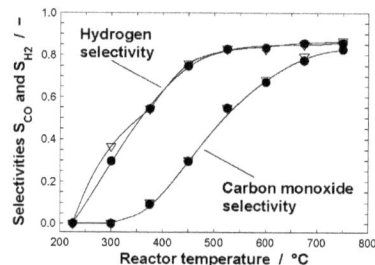
(b) Carbon monoxide and hydrogen selectivities $S_{CO}$ and $S_{H_2}$

Figure 3.6: Butane conversion $\eta$ and hydrogen yield $\Psi$ (a), carbon monoxide selectivity $S_{CO}$ and hydrogen selectivity $S_{H_2}$ (b) as functions of reactor temperature for packed bed reactors with a catalyst loading of 2.0 wt% Rh and $SiO_2$ fiber plugs in an Inconel reactor tube (•) and a quartz reactor tube ($\triangledown$), respectively, for an inlet C/O ratio $\phi = 0.8$. The solid lines through the data points are curve fits.

### 3.4.5 Temperature profile along the packed bed

For the results presented in this article, isothermal conditions along the packed bed were assumed. To prove that the temperature in the packed bed is identical with the nominal oven temperature and that the temperature variation along the packed bed is negligible, four thermocouples were inserted into the packed bed with about 4 mm distance between each other. The first thermocouple reached the ceramic fiber plug at the inlet of the packed bed. The second and third one were in the packed bed itself. The fourth measured the temperature in the fiber plug at the outlet. In Fig. 3.7, the four measured temperatures are shown for a typical measuring cycle. It is obvious that the four temperatures only differ very slightly from the nominal oven temperature. The temperature profiles in the packed bed are shown for the 8 points where the gas composition was measured, using the temperatures averaged over 5 minutes before each measurement point and referenced to the first thermocouple. For low oven temperatures, the maximum temperature was reached at the outlet of the packed bed, whereas for higher oven temperatures, the temperature was maximal inside the packed bed.

However, the highest temperature variation was always below 2.5 K, corresponding to a relative temperature deviation of less than 0.5%. Therefore, the assumption of isothermal conditions in the packed bed reactor was easily satisfied.

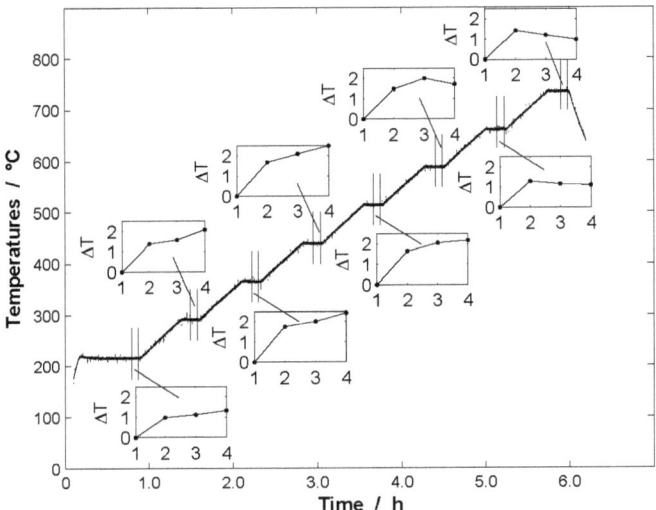

Figure 3.7: Temperatures of four thermocouples in the reactor (1: at the inlet, 2 and 3: in the packed bed, 4: at the outlet) during one measurement cycle. The small plots show the four temperatures averaged over 5 minutes before each measurement point and referenced to the first thermocouple.

## 3.5 Discussion

The use of $Rh/Ce_{0.5}Zr_{0.5}O_2$ nanoparticles as catalysts for the Partial Oxidation (POX) of butane allowed production of $H_2$ and CO at temperatures as low as 300°C. While maximum conversion and highest selectivity to $H_2$ and CO were found at high temperatures, a satisfactory catalytic performance could already be achieved in the present system at relatively low temperatures between 500 and 600°C. This temperature range was suggested for the incorporation in intermediate-temperature SOFC systems [18, 51].

The more detailed investigation of different reactor materials revealed that the ceramic fiber plugs had a pronounced influence on the overall performance. Fibers consisting of $Al_2O_3/SiO_2$ resulted in an enhanced butane conversion and $H_2$/CO production between 300 and 400°C if compared to pure $SiO_2$ fibers (Fig. 3.4). At 400°C, the performance of a $SiO_2$ fiber based

| Reactor temperature | 525°C | | 600°C | |
|---|---|---|---|---|
| Material of fiber plugs | $Al_2O_3/SiO_2$ | $SiO_2$ | $Al_2O_3/SiO_2$ | $SiO_2$ |
| Butane conversion $\eta$ | 54% | 72% | 65% | 81% |
| Hydrogen yield $\Psi$ | 31% | 55% | 45% | 72% |
| $H_2$ selectivity $S_{H_2}$ | 56% | 76% | 70% | 80% |
| CO selectivity $S_{CO}$ | 20% | 41% | 40% | 56% |
| $H_2$ mole fraction $X_{H_2}$ | 10% | 16% | 14% | 20% |
| CO mole fraction $X_{CO}$ | 3% | 7% | 6% | 10% |

Table 3.1: Comparison of performance for packed beds with $Al_2O_3/SiO_2$ and $SiO_2$ fiber plugs at 525°C and 600°C, respectively, using $Rh/Ce_{0.5}Zr_{0.5}O_2$ nanoparticles with 0.5 wt% Rh.

reactor started to exceed those of $Al_2O_3/SiO_2$ fibers and grew more evident around 500 to 650°C (Table 3.1). In order to better study this effect, we investigated the chemical reactions in the absence of any noble metal or ceria-based ceramics and used $SiO_2$-sand filled tubes with ceramic plugs (Fig. 3.4(a)). The use of alumina containing fibers significantly reduced the butane conversion by up to a factor of three at 600°C. A closer view on the corresponding very low amounts of produced $H_2$ and CO indicated that most of the butane is fully oxidized. This clearly illustrated the important contribution of non-noble metal catalyzed butane conversion and possible homogeneous reaction of butane at temperatures above 600°C. At such high temperature, appreciable amounts of butane were obviously reacting without the need of noble metal activation. Previous studies suggested that oxidative dehydrogenation and POX of different hydrocarbons are a combination of gas-phase homogeneous and catalyzed heterogeneous reactions [8, 29–36], as mentioned before. The low amount of produced hydrocarbon by-products shown in this study was suggested by [35] as a sign of strong interplay between homogeneous and heterogeneous reactions.

The Rh loading on $Ce_{0.5}Zr_{0.5}O_2$ nanoparticles had an important impact on the production of syngas. However, the Rh loading did not affect all reactions taking place in the reactor in the same way: While the butane conversion $\eta$ was remarkably higher for an Rh loading of 2.0 wt% compared to 0.5 wt%, the production of $H_2$ was almost identical above 600°C for both loadings (Fig. 3.5). While the $H_2$ production for 0.5 wt% Rh exceeded significantly the production of $H_2$ for 0.25 wt%, the difference in butane conversion and CO production was not large for these Rh loadings. These two effects may again be discussed together with results from reactors containing no Rh-based catalyst at all. Figures 3.4(a) and 3.5(a) clearly show that the reaction was taking off (light-up) at around 300°C resulting in a butane conversion independent of the Rh loading (Fig. 3.5(a)) or the type of fiber plugs (Fig. 3.4(a)). The apparent lack of any effect of increasing Rh catalyst loading can now be compared to the lack of CO selectivity and very low $H_2$ production at 300°C. Apparently, a light-off at 300°C resulted in full oxidation of butane.

The production of $H_2$ and CO, indicated by their selectivities, differed in their dependence on reactor temperature. For 2.0 wt% Rh loading and $SiO_2$ fiber plugs, the $H_2$ selectivity reached half of its maximum value already below 350°C, whereas the CO selectivity reached this point at around 500°C. The entire temperature dependent behavior showed much higher $H_2$ selectivity than CO selectivity. Only above 500 to 550°C, the $H_2$ and CO selectivities had similar values: for a reactor with 2.0 wt% Rh and 600°C, the value of $S_{H_2}$ was 83% and the value of $S_{CO}$ 67%. At 750°C, $S_{H_2}$ and $S_{CO}$ amounted to 86% and 83%, respectively. For lower temperatures, the selectivity towards CO was evidently lower compared to that towards $H_2$: at 450°C, $S_{H_2}$ was 40% and at 375°C, $S_{H_2}$ equaled only 17%. This strongly suggests that POX was not the only reaction producing $H_2$, since POX would have led to similar $H_2$ and CO selectivities. Instead, the reaction was dominated at lower temperatures (below 500°C) by Total Oxidation (TOX) of butane followed by $H_2O$ consuming reactions like Steam Reforming (SR) and Water Gas Shift (WGS). $H_2O$, which was produced by TOX beside $CO_2$, reacted with butane by SR to $H_2$ and CO, where part of this CO reacted subsequently with $H_2O$ by WGS to $CO_2$ and $H_2$. This combination of TOX, SR, and WGS explained the stronger $H_2$ production compared to the CO production, which could not be explained by POX alone. For higher temperatures, POX was the dominating reaction, where $H_2$ and CO were produced likewise. $CO_2$ producing reactions like TOX and WGS were unimportant for higher temperatures.

Compared to previous studies on hydrocarbon processors, the catalyst used for this study showed very high syngas production at relatively low temperature. An autothermal reforming path was investigated by Wang and Gorte for hydrocarbons on Pd/ceria [38], reporting a CO selectivity of 6% and butane conversion of 11% at 450°C using Pd/ceria as catalyst for SR of butane, compared to a maximum CO selectivity of 30% and butane conversion 71% at the same temperature in the present investigation. At 527°C, Wang and Gorte [52] achieved butane conversion of 41% by SR on Pd/ceria, in contrast to 92% butane conversion at 525°C in this study. Acharya et al. [53] investigated SR of isobutane on $Pt/CeO_2/Gd_2O_3$ and demonstrated butane conversion of 63% and hydrogen yield (as produced hydrogen per hydrogen input from butane and $H_2O$) of 39% at 600°C, compared to nearly complete butane conversion and 77% hydrogen yield in this investigation at the same temperature. Huff et al. [6] performed POX of butane over Pt coated foam monoliths with $\phi = 0.9$ resulting in butane conversion of 99%, hydrogen yield of 31%, and CO selectivity of 79% at 1041°C. In the present study, similar butane conversion and CO selectivity (100% and 77%, respectively) could be obtained at much lower temperature of 675°C and with a corresponding higher hydrogen yield of 86%. The present investigation on supported Rh catalysts with a very open morphology resulted in higher $H_2$ and CO yields at lower temperatures compared to other studies, which illustrates the favorable morphological and chemical properties of the aerosol derived ceria/zirconia nanoparticles as a support for Rh. For all measurements, neither significant cracking of butane nor formation of soot could be found, which corresponded well with results from Wang and Gorte [38] and Hilaire et al. [54].

## 3.6 Conclusions

The present work investigated the capability of rhodium doped ceria/zirconia nanoparticles as catalysts for the production of syngas from butane at 225 to 750°C. The main issue of this study was to develop a high catalytic performance for intermediate temperatures of 500 to 600°C. In this range, a packed bed reactor with catalyst nanoparticles of 2.0 wt% Rh loading and $SiO_2$-based plugs achieved nearly complete butane conversion with a hydrogen yield $\Psi$ of 77%. This resulted in an off-gas containing a $H_2$ mole fraction $X_{H_2}$ of 21% and a CO mole fraction $X_{CO}$ of 13%. In spite of its wide use as a sealing material or monolith base [6, 10, 26–28], the influence of $Al_2O_3$ in the sealing plugs of the reactor strongly affected the overall performance and suggested a more detailed investigation of non-noble metal catalyzed and homogeneous contributions to butane conversion in such reactors. Our results showed that the here-presented reactor configurations may well be suited to provide small and portable butane processing units for applications together with micro fuel cells.

# Chapter 4

# Disk-shaped packed bed micro-reactor

Parts of this chapter are published in:
N. Hotz, N. Osterwalder, W.J. Stark, N.R. Bieri, D. Poulikakos. Disk-shaped packed bed micro-reactor for butane-to-syngas processing, CHEMICAL ENGINEERING SCIENCE 63 (2008) p. 5193 - 5201.

## 4.1 Abstract

A novel disk-shaped packed bed micro-reactor containing Rh/ceria/zirconia nanoparticles is investigated with respect to catalytic butane-to-syngas processing at moderate temperatures of 550°C. The main goal of this study is the development of an efficient butane processor which can be integrated into a micro Solid Oxide Fuel Cell system due to its small size, easily packaged geometry in layered microdevices, high compactness, low pressure drop, and low reaction temperature. It is shown that Rh/ceria/zirconia has an excellent long-term stability and achieves very high $C_4H_{10}$ conversion and syngas selectivity, considering the relatively low operating temperature. The yields of $H_2$ and CO can be increased up to 71% and 57%, respectively, by optimizing operational parameters such as the C/O ratio and the total inlet flow rate. The introduced disk-shaped packed bed reactor shows significant advantages in catalytic behavior, at a 6.5 times lower pressure drop compared to an equivalent tubular packed bed reactor. This increased catalytic performance is pursued extensively by investigating possible reaction pathways in three regions of the radial-flow reactor, leading to the significant discovery of a threefold pathway of syngas production on a single catalyst. To this end, it is shown that the excellent selectivities to $H_2$ and CO for high flow rates are due to the combination of Partial Oxidation, Steam Reforming, and Dry Reforming of $C_4H_{10}$, indicating one direct and two indirect reaction paths. Finally, the effect of catalytically active surface area and catalyst site density on the activity is investigated, proving that both parameters have to be simultaneously taken into account to describe the reactor performance appropriately.

## 4.2 Introduction

Micro fuel cell systems generating electric power of the order of a few watts are seen to have promising potential to compete with conventional battery systems for small portable electronic devices due to their higher power density both per volume and per mass. Small fuel cell systems fed by hydrocarbons combine the high energetic efficiency of fuel cells with the high availability and easy storage of hydrocarbon fuels [20]. Modern materials for Solid Oxide Fuel Cells (SOFCs) lead to higher efficiencies compared to other types of fuel cells at intermediate operating temperatures in the range of 500 and 600°C [19] using syngas as fuel. An interesting hydrocarbon fuel widely available for this application is butane, allowing the production of a $H_2$- and CO-rich syngas with high efficiency and relatively easy storage in liquid phase at room temperature and low pressure.

The catalytic partial oxidation of butane was investigated in the past and the characteristics of different catalysts for this purpose were tested, e.g. [6, 7, 38, 52, 53, 55]. However, these studies do not account for many typical requirements of a fuel processor operated as part of an entire micro fuel cell system.

The novelty of this study is to introduce an easy to pack, disk-shaped packed bed micro-reactor for high-temperature fuel processors with a design feasible for practical applications such as micro SOFC systems presented previously [19]. Crucial requirements for the integration of a fuel processor into an entire micro SOFC system are easy integration into a layered device, a small reactor volume, a highly compact design, low pressure drop, and low reaction temperature. Since the geometry of planar SOFC membranes leads to disk-shaped fuel cell designs, a likewise disk-shaped reactor for the fuel processing increases the compactness of the entire system. The main goal of this study is to show the feasibility of a disk-shaped packed bed micro-reactor for butane-to-syngas processing as part of a micro SOFC system, achieving excellent catalytic activity and high long-term stability at a relatively low operating temperature and pressure drop within a compact and small reactor.

As a noble metal, rhodium was chosen for its excellent performance in butane-to-syngas conversion [55]. The advantageous use of ceria/zirconia as a catalyst support for high-temperature reactions was demonstrated earlier [47, 49].

A desirable reaction path for syngas production from hydrocarbons is Partial Oxidation (POX), written as

$$C_4H_{10} + 2\,O_2 \rightarrow 5\,H_2 + 4\,CO, \tag{4.1}$$

which achieves high yields of $H_2$ and CO. POX has been identified as the ideal reaction path for hydrocarbon processing to syngas and investigated experimentally and numerically for different hydrocarbon fuels, e.g., methane [42, 56, 57] and higher alkanes [58, 59].

Another effective reaction for hydrocarbon processing is Steam Reforming (SR), where butane reacts with water:

$$C_4H_{10} + 4\,H_2O \rightarrow 9\,H_2 + 4\,CO. \tag{4.2}$$

A second water consuming reaction which might take place in a micro-reactor besides SR is Water Gas Shift (WGS):

$$CO + H_2O \rightarrow CO_2 + H_2. \tag{4.3}$$

Syngas can be produced from hydrocarbons by $CO_2$ or Dry Reforming (DR):

$$C_4H_{10} + 4\,CO_2 \rightarrow 5\,H_2 + 8\,CO. \tag{4.4}$$

A well performing butane processor should show high selectivity towards $H_2$ and CO instead of Total Oxidation (TOX) products of butane, which is written as

$$C_4H_{10} + 6.5\,O_2 \rightarrow 5\,H_2O + 4\,CO_2, \tag{4.5}$$

and low methane production by decomposition of $C_4H_{10}$ or methanation,

$$CO + 3\,H_2 \rightarrow CH_4 + H_2O. \tag{4.6}$$

## 4.3 Experiments

### 4.3.1 Catalyst preparation

$Ce_{0.5}Zr_{0.5}O_2$ nanoparticles with optional rhodium doping were prepared in a one-step process by flame spray synthesis described in previous studies [40, 47, 49, 55]. For the ceria/zirconia precursor, cerium(III) 2-ethylhexanoate (12 wt% Ce, Shepherd Chemical Company) and zirconium(IV) 2-ethylhexanoate (18 wt% Zr, Borchers GmbH) were mixed to result in a metal molar ratio Ce/Zr of 1:1 and diluted with xylene to a total metal concentration of 0.8 mol $L^{-1}$. Rhodium(III) 2-ethylhexanoate (UMICORE AG & Co.) was added to the Ce/Zr-precursor such that the calculated rhodium content in the ternary system Rh/ceria/zirconia (Rh/$Ce_{0.5}Zr_{0.5}O_2$) was 2.0, 0.5, and 0.25 wt%.

## 4.3.2 Catalyst characterization

The specific surface area of the catalyst was measured using nitrogen adsorption on a Tristar (Micromeritics Instruments) at 77 K with the BET method and used to calculate the mean particle diameter. The phase composition and formation of ceria/zirconia mixed oxides was determined by X-ray powder diffraction on a Stowe STADI-P2 (Ge monochromator, Cu $K_{\alpha 1}$, PSD detector). The catalytic nanoparticles were analyzed by flame atomic absorption spectrometry (AAS) on a Varian SpectrAA 220FS. The chemisorption of the Rh surface was measured using an ASAP 2010 (Micromeritics Instruments). The samples were reduced at 400°C for 90 minutes using pure hydrogen and cooled down to 40°C under helium atmosphere. From the results of hydrogen chemisorption measurements, the metal dispersion was calculated as molecular hydrogen adsorbed dissociatively on Rh metal (H/Rh = 1) [60]. The active Rh site density $\rho_{Rh,s}$ was calculated as active Rh sites $n_{Rh,s}$ per surface area of the Rh/ceria/zirconia catalyst.

Transmission electron micrographs (TEM) of fresh and spent catalyst particles were recorded on a CM30 ST (Philips, 300 kV voltage, point to point resolution 0.19 nm) equipped with an energy dispersive X-ray spectrometer (EDX) to analyze the chemical composition of the particles qualitatively. For the TEM and EDX analysis, all samples were dispersed in ethanol. The spent catalyst material was ground manually to separate the Rh/ceria/zirconia nanoparticles from larger silica particles used in the reactors. All results shown later were obtained from samples prepared in an ultrasonic bath and deposited onto a carbon-coated TEM grid.

## 4.3.3 Reactor

It has been previously shown that the catalyst described above can be used for the production of syngas from butane in small tubular reactors (i.d. 2 mm) [55]. However, the geometry of tubular reactors contradicts to the requirement of compact design in micro fuel cell systems, especially if in context with disk-shaped SOFC membranes. Therefore, we used a disk-shaped design for the micro-reactor with a catalytically active volume of 10 mm diameter and 0.5 mm height. In addition to the compactness of the geometry, this design led to lower pressure drop due to the increasing flow cross section along the reaction path, as it has been shown for low-temperature fuel processors [61], and higher catalytic activity.

The air/butane mixture entered the reactor from the left side at the center of the reactor and flowed radially outwards through the catalytically active packed bed consisting of Rh/ceria/zirconia nanoparticles and $SiO_2$ sand, as shown in Fig. 4.1. The reactor walls consisted of quartz glass and the packed bed was fixed on both sides by quartz glass wool.

The packed bed consisted of nanoparticles and $SiO_2$ sand (average diameter: 200 $\mu$m) of different mass ratios: for the standard configuration, 10 mg Rh/ceria/zirconia nanoparticles with

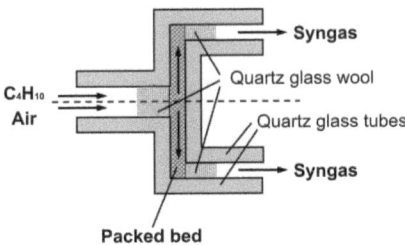

Figure 4.1: Schematic of the disk-shaped packed bed reactor

an Rh loading of 2.0 wt% were mixed with 30 mg $SiO_2$ sand. To investigate the different reactions taking place along the radial flow direction, disk-shaped packed bed reactors with 6 and 8 mm diameter were tested. Since all three reactors contained catalyst with identical volumetric density, the reactors of 6, 8, and 10 mm diameter were identical except for their radial path length. By subtraction of the molar gas flow of these three reactors, the molar production rate of different gas species within three regions could be calculated: Region 1 up to 3 mm radial path length, Region 2 between 3 and 4 mm, and Region 3 between 4 and 5 mm radius.

To investigate the effect of Rh loading and surface-to-volume ratio for constant total amount of Rh, 40 mg Rh/ceria/zirconia (0.5 wt% Rh) and 20 mg $SiO_2$ were tested as well as 80 mg Rh/ceria/zirconia (0.25 wt% Rh) and 6 mg $SiO_2$. The amount of $SiO_2$ sand was adjusted to keep the volume of solids in the packed bed constant.

### 4.3.4 Test setup

Expanded butane from a liquid tank (PanGas, 3.5, 99.95%) at 2.5 bar was mixed with compressed synthetic air (79% $N_2$, 21% $O_2$, PanGas, 5.6, purity of both species: 99.9999%) from a gas tank. Both flow rates were controlled by Low Delta-P flow meters (Bronckhorst), allowing operation of the reactor slightly above the ambient pressure. The butane/air mixture was fed into the disk-shaped reactor heated by a tube furnace (MTF 12/38/250, Carbolite). The reactor tubes were heated over a length of 30 cm, ensuring isothermal conditions inside the packed bed reactor placed in the middle of the furnace. The product gas leaving the furnace was maintained at around 115°C to avoid condensation of $H_2O$. The gas composition was analyzed by a gas chromatograph (6890 GC) coupled with a mass spectrometer (5975 MS, Agilent), using a HP-MOLSIV and a HP-PlotQ column (Agilent), respectively. Helium (PanGas, 5.6, 99.9996%) was added as an internal standard for GC calibration. Under typical run conditions, the molar product gas balances of C, H, and O were closed within 5%.

## 4.3.5 Testing procedure

The reactor was heated up to 550°C at a heating rate of 12.5°C/min. During the last 10 min of the heating process, the reactor was flushed with 20 sccm air. At 550°C, the reactor was fed with a butane/air mixture and eight measurements lasting about 44 min each were performed. After the measurements the reactor was flushed again with 20 sccm air for 1 hour while cooling down to about 350°C. The reactor was operating under permanent butane/air flow at 550°C for totally 345 min per measurement cycle.

Subsequently, experimental results were analyzed using characteristic values quantifying the catalytic behavior of the packed bed reactor, based on the mole fractions of the outlet gas measured by the GC/MS. The butane conversion $\eta$ of the micro-reactor was determined as the molar ratio between converted butane and inlet butane:

$$\eta = \frac{\dot{n}_{C_4H_{10},in} - \dot{n}_{C_4H_{10},out}}{\dot{n}_{C_4H_{10},in}}. \tag{4.7}$$

The selectivities for $H_2$ and CO read:

$$S_{H_2/CO} = \frac{\dot{n}_{H_2/CO,out}}{\dot{n}_{H_2/CO,out} + \dot{n}_{H_2O/CO_2,out}}. \tag{4.8}$$

The yields of $H_2$, $H_2O$, CO, and $CO_2$ were defined as the molar ratio of these species in the outlet gas and the molar amount of $H_2$ or C in the inlet gas in form of $C_4H_{10}$:

$$Y_{H_2/H_2O} = \frac{\dot{n}_{H_2/H_2O,out}}{5 \cdot \dot{n}_{C_4H_{10},in}} \tag{4.9}$$

and

$$Y_{CO/CO_2} = \frac{\dot{n}_{CO/CO_2,out}}{4 \cdot \dot{n}_{C_4H_{10},in}}. \tag{4.10}$$

The yield of $CH_4$ was calculated as a function of the molar ratio of $CH_4$ in the outlet gas and $C_4H_{10}$ in the inlet gas referenced to their content of $H_2$:

$$Y_{CH_4} = \frac{2 \cdot \dot{n}_{CH_4,out}}{5 \cdot \dot{n}_{C_4H_{10},in}}. \tag{4.11}$$

Each operating condition was defined by the C/O ratio or equivalence ratio $\phi$ based on POX, calculated as

$$\phi = \frac{2 \cdot \dot{n}_{C_4H_{10},in}}{\dot{n}_{O_2,in}}, \qquad (4.12)$$

and the total inlet flow rate $\dot{V}_{gas,in}$. This flow rate was varied from 10 to 30 sccm, leading to Gas Space Velocities, defined as the ratio of total volumetric flow rate at the reactor inlet and reactor volume, between 11.64 and 34.93 s$^{-1}$ and Space Times, the reciprocal of the Gas Space Velocity, in the range of 85.9 and 28.6 ms.

## 4.4 Results

### 4.4.1 Reactor composition and catalyst characterization

For the standard reactor configuration, 10 mg Rh/ceria/zirconia nanoparticles with a Rh loading of 2.0 wt% were mixed with 30 mg of SiO$_2$ sand. The average diameter of the nanoparticles was calculated as 10.4 nm using the BET specific surface area of 90 m$^2$/g. The Rh dispersion measured by chemisorption amounted to 25.3%, indicating a number of active Rh surface sites of 0.493 $\mu$mol on a catalytically active surface area of 0.90 m$^2$.

Three reactor configurations were tested, as shown in Table 4.1, using Rh/ceria/zirconia nanoparticles with different nominal Rh loadings of 2.0, 0.5, and 0.25 wt%. The nominal mass of Rh was 0.2 mg for all three configurations and the reactor volume was kept constant by mixing 10 mg Rh/ceria/zirconia catalyst with 30 mg SiO$_2$ sand (DSPB-10), 40 mg catalyst with 20 mg SiO$_2$ sand (DSPB-40), and 80 mg catalyst with 6 mg SiO$_2$ (DSPB-80). Since the Rh dispersion measured by chemisorption changed between 25.3 and 49.5%, the active Rh surface area and the number of active Rh surface sites $n_{Rh,s}$ was highest for the DSPB-40 reactor and lowest for the DSPB-10 reactor. However, the active Rh site density per total Rh/ceria/zirconia surface area $\rho_{Rh,s}$ was highest for the DSPB-10 reactor (0.547 $\mu$mol/m$^2$) and lowest for the DSPB-80 (0.094 $\mu$mol/m$^2$).

Transmission electron microscopic images of the fresh catalyst nanoparticles showed crystalline (recognizable from the nicely facetted surfaces), highly regular nanoparticles of around 10 nm diameter (Fig. 4.2(a) and 4.2(b)). The morphology and crystalline structure did not change during catalytic operation, as shown by TEM images of catalyst used for 40 h (Fig. 4.2(c) and 4.2(d)). It has to be noted that no carbon deposition was seen on any spent catalyst, the crystalline particles always showing clean surfaces. Also the analysis by EDX did not suggest any change in composition of the catalyst before and after usage. There is no indication of the presence of any carbon formation on the catalyst.

|  | DSPB-10 | DSPB-40 | DSPB-80 |
|---|---|---|---|
| Catalyst mass [mg] | 10.0 | 40.0 | 80.0 |
| Rh loading [wt%] | 2.00 | 0.50 | 0.25 |
| Rh mass [mg] | 0.2 | 0.2 | 0.2 |
| $SSA_{BET}$ [m$^2$/g] | 90 | 109 | 107 |
| $d_{BET}$ [nm] | 10.4 | 8.6 | 8.8 |
| Rh/Ce$_{0.5}$Zr$_{0.5}$O$_2$ surface [m$^2$] | 0.90 | 4.36 | 8.56 |
| Rh surface [m$^2$] | 0.022 | 0.043 | 0.036 |
| Rh dispersion [%] | 25.3 | 49.5 | 41.2 |
| $n_{Rh,s}$ [μmol] | 0.493 | 0.962 | 0.802 |
| $\rho_{Rh,s}$ [μmol/m$^2$] | 0.547 | 0.221 | 0.094 |

Table 4.1: Properties of the catalysts used in three tested disk-shaped reactors

### 4.4.2 Stability test for a disk-shaped reactor

A disk-shaped reactor containing 10 mg Rh/ceria/zirconia catalyst with a Rh loading of 2.0 wt% was tested for 7 measurement cycles described before under 20 sccm butane/air mixture and a C/O ratio $\phi = 0.8$ to investigate the stability of the catalyst.

During the first two measurement cycles, the butane conversion decreased and the selectivities towards H$_2$ and CO were relatively low, as shown in Fig. 4.3. This was due to the fact that the catalyst was used without any pretreatment to reduce the nanoparticles by contact with a H$_2$-rich gas flow, as usually done for such catalysts, e.g. [7, 8, 34, 62]. One advantage of the herein-used catalytic nanoparticles was that no such pretreatment was necessary, since the catalytic activity of the Rh/ceria/zirconia was strong enough from the beginning to convert C$_4$H$_{10}$ to H$_2$ and CO. After these two days of leveling, the catalytic performance was very stable for at least the next 5 cycles, meaning 40 measurements during 29 h of operation. No tendency towards degradation of the catalyst or carbon deposition during this time could be seen. Therefore, this procedure of stabilizing the catalyst for 2 measurement cycles with 20 sccm butane/air mixture and $\phi = 0.8$ and subsequent measuring under varying operational parameters for 5 measurement cycles was used for all following investigations.

### 4.4.3 Effect of C/O ratio and total inlet flow rate on catalytic performance

The effect of the C/O ratio $\phi$ and the total inlet flow rate $\dot{V}_{gas,in}$ was investigated for a disk-shaped reactor containing 10 mg Rh/ceria/zirconia catalyst with 2.0 wt% Rh. The C/O ratio was varied from 0.5 to 1.2 and the total inlet flow rate $\dot{V}_{gas,in}$ was changed from 10 to 30 sccm, resulting in space times of 85.9 to 28.6 ms. Figure 4.4(a) shows that the butane conversion decreased with increasing C/O ratio and increasing total inlet flow rate. For all flow rates,

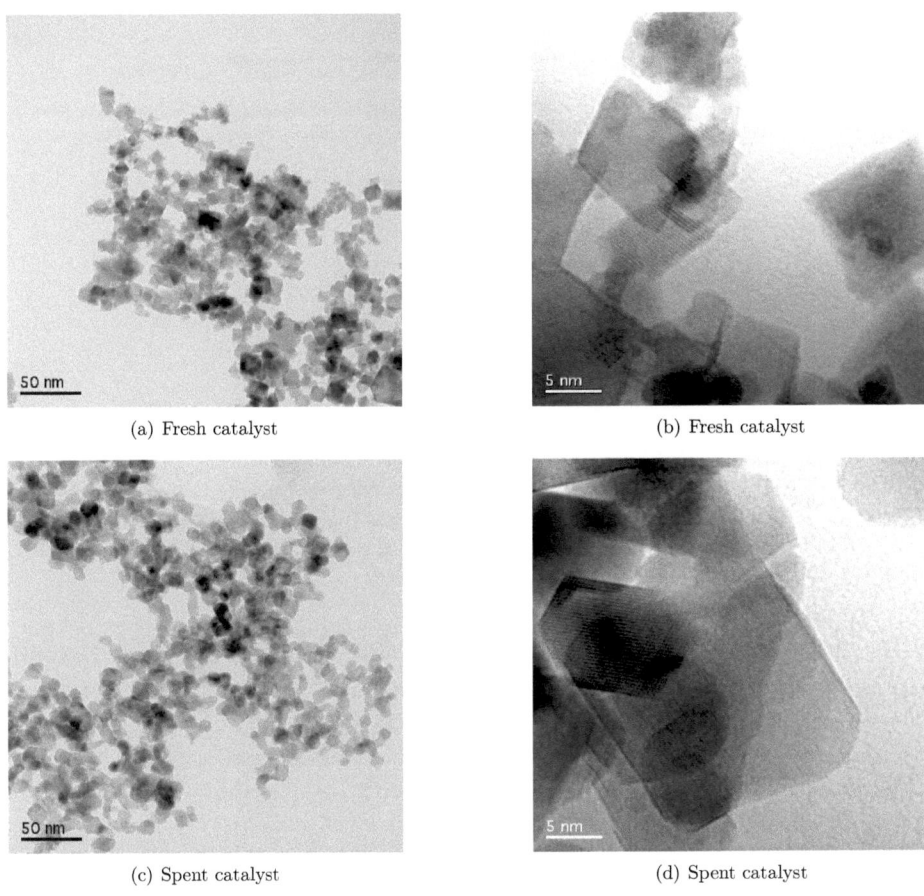

Figure 4.2: Transmission electron micrographs of 2.0 wt% $Rh/Ce_{0.5}Zr_{0.5}O_2$ nanoparticles (a) and (b) before use and (c) and (d) after 40 h of catalytic operation under butane/air flow at 550°C.

the butane conversion was above 89.9% for C/O ratios below 0.9. For C/O ratios above 0.9, the butane conversion dropped much faster for flow rates higher than 10 sccm and there was only little difference in butane conversion for flow rates between 15 and 30 sccm. $H_2$ and CO selectivity increased with increasing C/O ratio and increasing total inlet flow rate. By increasing the flow rate from 10 to 15 sccm and further, both selectivities could be improved significantly. However, the difference in selectivity between 20 and 30 sccm was rather small, especially for high C/O ratios, where all flow rates converged towards the $H_2$ selectivity of 88.6% and the CO selectivity of 78.0% reached at $\phi = 1.2$.

Considering the opposite effect of increasing $H_2$ selectivity and decreasing butane conversion

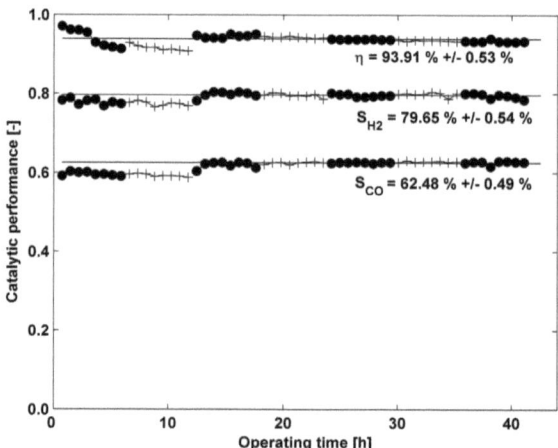

Figure 4.3: Disk-shaped reactor: Catalytic performance (butane conversion $\eta$, $H_2$ selectivity $S_{H_2}$, and CO selectivity $S_{CO}$) is shown for 7 measurement cycles, lasting 5.8 h each, for a C/O ratio $\phi = 0.8$ and a total inlet flow rate $\dot{V}_{gas,in} = 20$ sccm.

for increasing C/O ratios, the $H_2$ yield reached maximal values at around $\phi = 0.7$, though showing small variation in $H_2$ yield for C/O ratio between 0.6 and 1.0 and inlet flow rate of 20 sccm and higher (Fig. 4.4(b)). The highest $H_2$ yield was reached at $\phi = 0.7$ and 30 sccm amounting to 70.8%. The $H_2O$ yield decreased with increasing C/O ratio with very little effect of the flow rate, especially for higher C/O ratios. The $CH_4$ yield reached a maximum at $\phi = 1.0$ for 10 sccm with $Y_{CH_4} = 17.4\%$. For flow rates above 15 sccm, the $CH_4$ yield was much lower, reaching maxima of $Y_{CH_4}$ below 10.5%. The CO yield increased similarly to the $H_2$ yield with increasing total inlet flow rate, with maximal values at $\phi = 0.9$ for higher flow rates ($Y_{CO} = 56.6\%$ for 30 sccm), at $\phi = 1.0$ for 15 sccm ($Y_{CO} = 54.0\%$), and at $\phi = 1.1$ for the lowest flow rate of 10 sccm ($Y_{CO} = 50.6\%$). For lower C/O ratios, increasing the flow rate from 10 to 20 sccm had a significant positive effect on the CO yield. The $CO_2$ yield decreased with increasing C/O ratio and increasing total inlet flow rate with relatively little effect of the flow rate on the $CO_2$ yield.

For all measurements (as well as for all other measurements presented in this study), no significant amount of lower hydrocarbon such as propane, ethane, propylene, or ethylene could be detected neither by the GC nor by the MS, indicating that cracking of $C_4H_{10}$ was negligible for these reactors. Similarly, no significant carbon deposition could be detected during all measurements, which could be expected by the excellent stability of the catalyst and the presented TEM images of spent catalyst. The product gas always contained only traces of $O_2$, implying practically full $O_2$ conversion. A reactor containing only silica sand and fiber plugs as

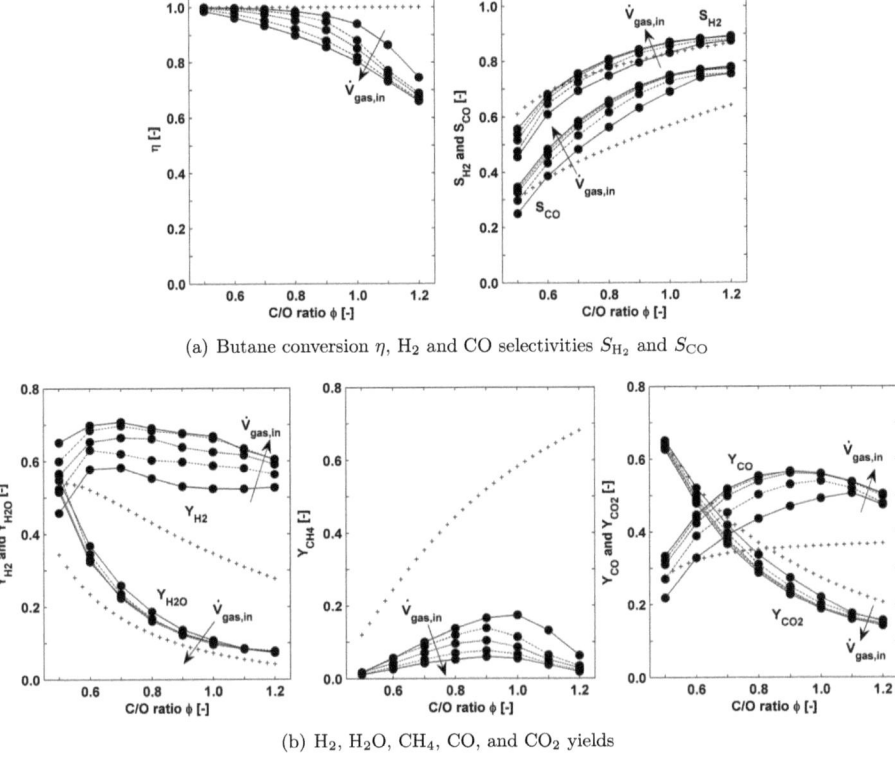

Figure 4.4: (a) Disk-shaped reactor: Butane conversion $\eta$, $H_2$ selectivity $S_{H_2}$, and CO selectivity $S_{CO}$ and (b) $H_2$, $H_2O$, $CH_4$, CO, and $CO_2$ yields are shown for C/O ratios $\phi$ between 0.5 and 1.2 and total inlet flow rates $\dot{V}_{gas,in}$ of 10, 15, 20, 25, and 30 sccm. + indicates the equilibrium state.

a reference measurement did not show any $H_2$ or CO production (not shown here). A very small conversion of $C_4H_{10}$ and $O_2$ and small amounts of $CO_2$ and $H_2O$ in the product gas proved the occurrence of some TOX but no POX, SR, or DR.

To avoid any influence of the thermocouples on the gas flow and the catalytic activity of the packed bed reactors, no thermocouples were used inside the reactors for all measurements presented here. However, to confirm the assumption of practically isothermal conditions inside the reactors, measurements with disk-shaped reactors with four thermocouples inserted into the inlet glass tube were conducted. The last thermocouple was right in front of the packed bed, the three others in the upstream direction with about 2 mm distance from each other. By increasing the inlet flow rate from 10 to 30 sccm, the thermocouple closest to the packed bed showed a temperature increase of only 3.3 K. The temperature difference between the thermocouple closest to the packed bed and the most upstream one changed from 5.3 to 6.3 K with increasing inlet flow rate. Another set of measurements was conducted with thermocouples in direct contact with the inlet and the outlet surface of the packed beds of 6, 8, and 10 mm diameter. The temperature right at the inlet surface of the packed bed, where the highest temperature increase has to be expected due to oxidative reactions, rose up to 563°C, indicating a temperature difference of about 13 K compared to the nominal reactor temperature of 550°C and therefore leading to a relative error of 1.6%. The temperature on the surface of the packed bed at the outlet did not change at all for varying flow rates and was independent of the reactor volume. The largest measured relative temperature difference was therefore below 2% for all tested flow rates, corroborating the assumption of practically isothermal reactors, as it has been shown for tubular reactors before [55].

### 4.4.4 Comparison of a disk-shaped and a tubular reactor

The disk-shaped reactor presented before was compared with a tubular packed bed reactor of the same reactor volume and catalyst loading (10 mg Rh/ceria/zirconia with 2.0 wt% Rh and 30 mg $SiO_2$ sand). The packed bed was placed at the inlet tube of the disk-shaped reactor (i.d. 2 mm) shown in Fig. 4.1 and was fixed with $SiO_2$ fibers. Except for the different geometry, both reactors were totally identical and were tested under the same conditions. The resulting catalytic performance is shown in Fig. 4.5, indicating significantly lower butane conversion and $H_2$ and CO selectivity for most operating points compared to the disk-shaped reactor. Except for the lowest inlet flow rate of 10 sccm and C/O ratios below 1.0, the butane conversion of the tubular reactor was below the values of the disk-shaped one. This difference was very pronounced for C/O ratios of 1.0 and higher, since the butane conversion of the tubular reactor experienced a stronger decay already at lower C/O ratios. The selectivities towards $H_2$ and CO were lower for the tubular reactor, especially for higher flow rates and higher C/O ratios. In contrast to the disk-shaped geometry, the selectivities decreased with increasing flow rate for high C/O ratios. Only at $\phi = 0.5$, both selectivities increased with increasing flow rate. For flow rates above 15 sccm, the $H_2$ and CO selectivities practically stagnated for C/O ratios higher than 0.8.

From Table 4.2, it can be seen that the maximal yields of $H_2$ and CO were significantly lower for the tubular reactor. The relative difference between both reactor designs increased up to 10% in $H_2$ yield and even 38% in CO yield for 30 sccm.

Besides the advantage of the disk-shaped geometry in the sense of catalytic activity, the disk-shaped packed bed resulted in a much lower pressure drop than the equivalent tubular reactor, which is clearly an essential aspect of micro fuel processors. In Table 4.2, the pressure drop of the entire test rig is shown for a C/O ratio $\phi = 0.8$. For all inlet flow rates from 10 to 30 sccm, the pressure drop was about 6.5 times lower for disk-shaped reactor. A similar reduction of the

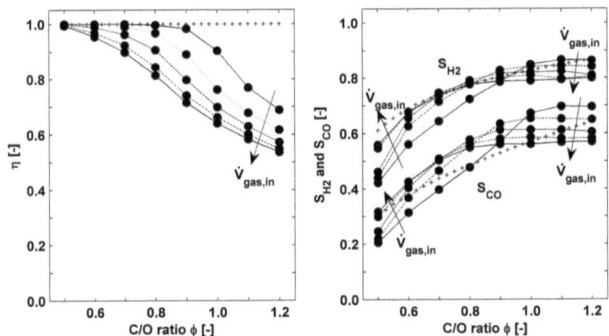

Figure 4.5: Tubular reactor TPB: Butane conversion $\eta$, $H_2$ selectivity $S_{H_2}$, and CO selectivity $S_{CO}$ are shown for C/O ratios $\phi$ between 0.5 and 1.2 and total inlet flow rates $\dot{V}_{gas,in}$ of 10, 15, 20, 25, and 30 sccm. + indicates the equilibrium state.

| Flow rate | | 10 | 15 | 20 | 25 | 30 | sccm |
|---|---|---|---|---|---|---|---|
| Max. $Y_{H_2}$ | DSPB-10 | 58.3 | 63.1 | 66.5 | 69.7 | 70.8 | % |
| | TPB | 54.6 | 59.5 | 61.3 | 63.4 | 64.4 | % |
| Max. $Y_{CO}$ | DSPB-10 | 50.6 | 54.0 | 56.2 | 56.6 | 56.6 | % |
| | TPB | 46.4 | 46.4 | 44.8 | 43.1 | 41.1 | % |
| Pressure drop | DSPB-10 | 30.3 | 41.3 | 49.2 | 57.0 | 61.4 | mbar |
| | TPB | 204.1 | 268.7 | 321.9 | 363.6 | 401.9 | mbar |

Table 4.2: Maximal $H_2$ and CO yields and pressure drop of the entire test rig for total inlet flow rates $\dot{V}_{gas,in}$ of 10, 15, 20, 25, and 30 sccm, comparing DSPB-10 and TPB.

pressure drop by using disk-shaped packed beds was already achieved [61], however, for much lower reactor temperatures and $\mu$m-sized catalyst particles.

### 4.4.5 Catalytic reactions taking place in different reactor regions

To investigate the reaction pathways taking place in a disk-shaped reactor rendering it more favorable than a tubular reactor and allowing a high catalytic performance especially for larger flow rates, the production and consumption of all gas species was measured by dividing the reactor in three regions: Region 1 from the inlet to a radius of 3 mm was investigated by testing a 6 mm diameter reactor, Region 2 between a radius of 3 and 4 mm by subtracting the results of reactors with 8 and 6 mm diameter, and Region 3 by subtracting the molar flow rates generated by reactors of 10 and 8 mm diameter. The molar flow rates of all detected gas species are presented in Fig. 4.6 for total inlet flow rates of 10, 20, and 30 sccm. For the lowest flow rate of 10 sccm, practically all $H_2$ was produced within Region 1 (within a disk of 6 mm diameter) and Regions 2 and 3 have very little effect on the $H_2$ production. Similarly, most

CO was generated in Region 1 at this flow rate, but especially at higher C/O ratios some CO was produced in Region 2 and significant amounts of CO in Region 3. When the flow rate was increased to 20 sccm, the importance of Region 3 was increased where the production of $H_2$ and CO was considerable, although Region 1 was still the dominant region of syngas production. At the highest flow rate of 30 sccm, all regions contributed to the generation of $H_2$ and CO and compensated the stagnation in syngas production from 20 to 30 sccm in Region 1. Accordingly, most of the butane conversion took place in Region 1 for all flow rates, however, for higher flow rates and C/O ratios (indicating increasing inlet flow of butane) the consumption of butane became more important in Region 2 and especially Region 3. It has to be noted that for all operating conditions the inlet oxygen was completely consumed within Region 1, meaning that no oxidative reactions such as POX and TOX could occur in Regions 2 and 3.

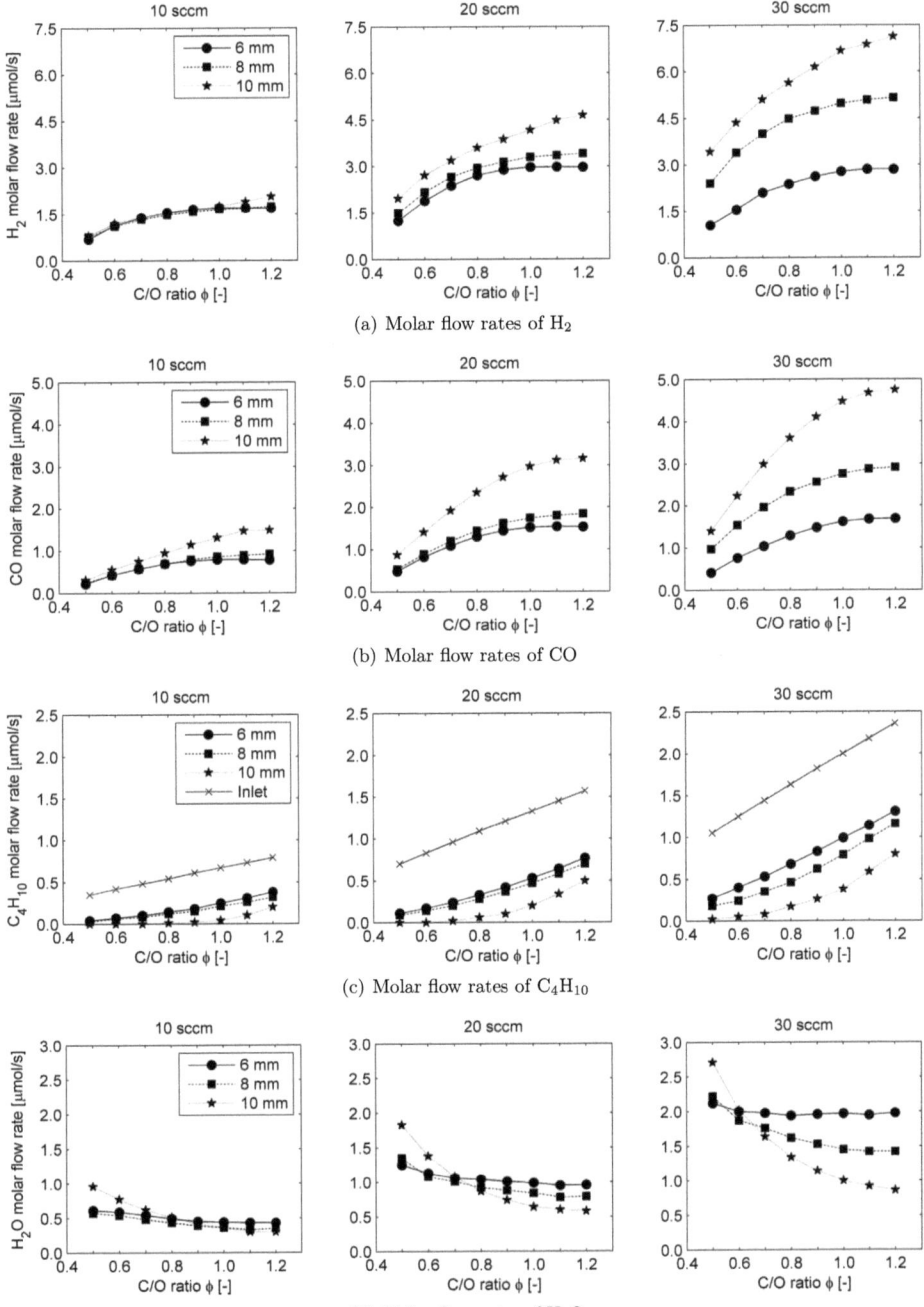

(a) Molar flow rates of $H_2$

(b) Molar flow rates of CO

(c) Molar flow rates of $C_4H_{10}$

(d) Molar flow rates of $H_2O$

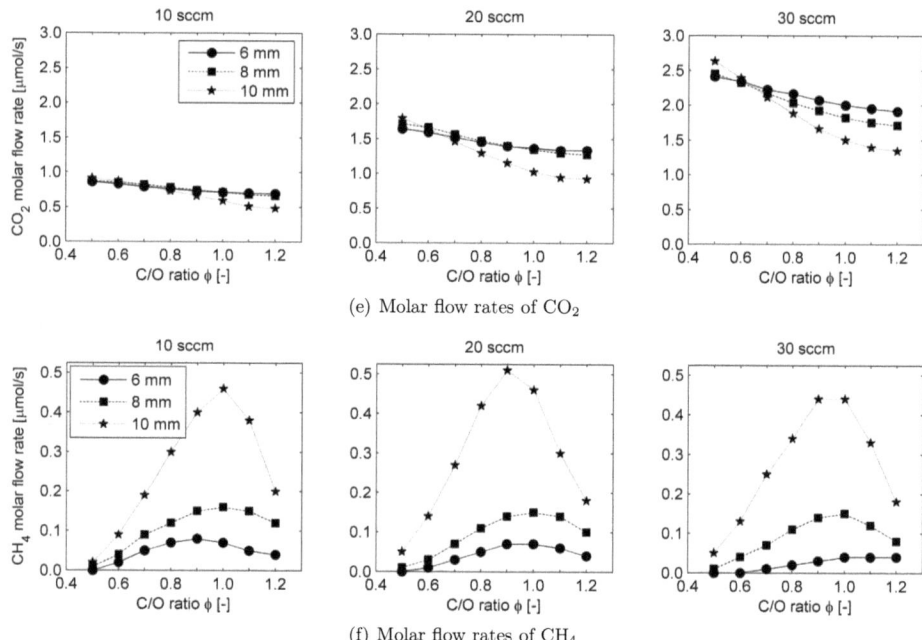

Figure 4.6: Molar flow rates of $H_2$, CO, $C_4H_{10}$, $H_2O$, $CO_2$, and $CH_4$ are shown for C/O ratios $\phi$ between 0.5 and 1.2 and total inlet flow rates $\dot{V}_{gas,in}$ of 10, 20, and 30 sccm after passing the regions of 6, 8, and 10 mm diameter.

If we continue our analysis with the undesired reaction products $H_2O$ and $CO_2$, it can be explained how production of $H_2$ and $CO_2$ could occur without $O_2$ in Regions 2 and 3: For higher flow rates and C/O ratios beyond 0.7, the consumption of $H_2O$ and $CO_2$ was remarkable. For example at 30 sccm, more than half of the $H_2O$ produced in Region 1 was consumed in Region 2 and 3 for C/O ratios above 1.0. Interestingly, next to $H_2O$ consuming reactions obviously occurring, the amount of consumed $CO_2$ was significant in Region 3 for higher flow rates and in Region 2 for 30 sccm. The production of $CH_4$ resulted mainly from reactions in Region 3, where the absolute amount of $CH_4$ was practically identical for all flow rates with maximum of 0.5 $\mu$mol/s at around $\phi = 1.0$. The accentuated production of $CH_4$ in Region 3 explained partly the increased consumption of $C_4H_{10}$ in this region of the reactor. For larger flow rates, the production of $CH_4$ was very low compared to all other product species, as shown before by the yield $Y_{CH_4}$.

### 4.4.6 Effect of catalytic surface area and catalyst site density on catalytic performance

The effect of the absolute number of catalytically active surface sites $n_{Rh,s}$ and the surface site density $\rho_{Rh,s}$ on the catalytic performance was investigated for three disk-shaped reactors with different catalyst compositions, as presented in Table 4.1, and shown for inlet flow rate of 10, 20, and 30 sccm. For 10 sccm, the $C_4H_{10}$ conversion increased with increasing surface site density, as shown in Fig. 4.7. The conversion was highest for the DSPB-10 and lowest for the DSPB-80, especially remarkable at low C/O ratios. The $H_2$ and CO selectivity was more dependent on the total number of catalytically active surface sites: They were highest for the DSPB-40 and lowest for the DSPB-80, although the difference to the DSPB-80 was less distinct with respect to the $C_4H_{10}$ conversion.

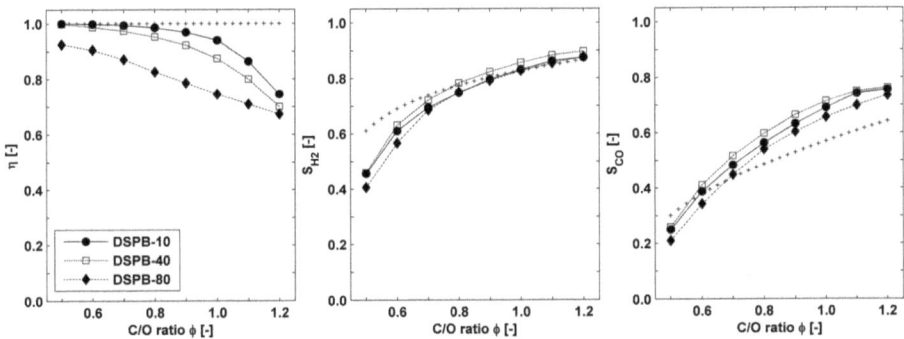

Figure 4.7: Disk-shaped reactors DSPB-10, DSPB-40, and DSPB-80: Butane conversion $\eta$, $H_2$ selectivity $S_{H_2}$, and CO selectivity $S_{CO}$ are shown for C/O ratios $\phi$ between 0.5 and 1.2 and a total inlet flow $\dot{V}_{gas,in}$ of 10 sccm, depending on the catalyst loading. + indicates the equilibrium state.

For higher inlet flow rates (Fig. 4.8 and 4.9), the difference between the DSPB-10 and DSPB-40 decreased: The $H_2$ and CO selectivities were slightly higher for the DSPB-40, whereas the $C_4H_{10}$ conversion was higher for the DSPB-10, except for C/O ratios above 1.1. However, the catalytic performance of the DSPB-80 decreased with increasing inlet flow rate, creating a larger difference of the $H_2$ and CO selectivities between DSPB-80 on one side and DSPB-10 and DSPB-40 on the other side, since the selectivities increased slightly for higher flow rates for DSPB-10 (as shown in subsection 4.4.3) and DSPB-40 and decreased for the DSPB-80.

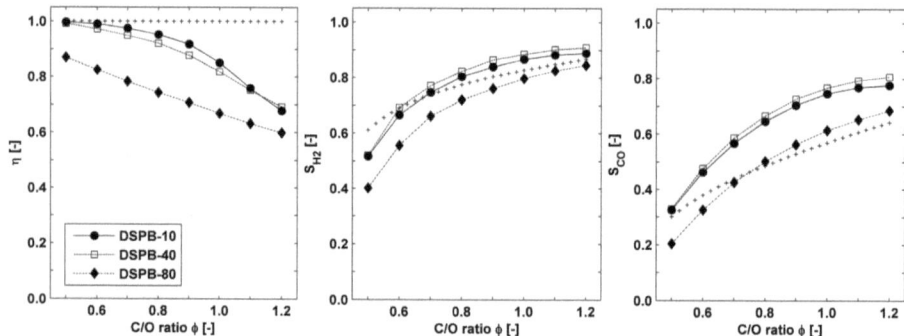

Figure 4.8: Disk-shaped reactors DSPB-10, DSPB-40, and DSPB-80: Butane conversion $\eta$, $H_2$ selectivity $S_{H_2}$, and CO selectivity $S_{CO}$ are shown for C/O ratios $\phi$ between 0.5 and 1.2 and a total inlet flow $\dot{V}_{gas,in}$ of 20 sccm, depending on the catalyst loading. + indicates the equilibrium state.

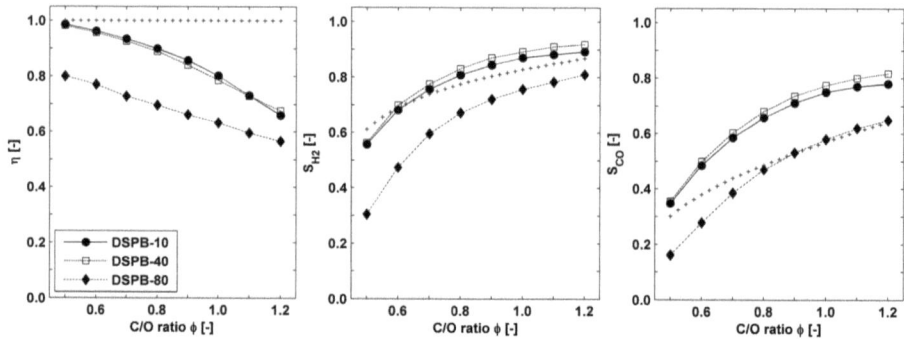

Figure 4.9: Disk-shaped reactors DSPB-10, DSPB-40, and DSPB-80: Butane conversion $\eta$, $H_2$ selectivity $S_{H_2}$, and CO selectivity $S_{CO}$ are shown for C/O ratios $\phi$ between 0.5 and 1.2 and a total inlet flow $\dot{V}_{gas,in}$ of 30 sccm, depending on the catalyst loading. + indicates the equilibrium state.

## 4.5 Discussion

### 4.5.1 Stability test

The results characterizing the catalytic activity of the tested Rh/ceria/zirconia nanoparticles showed excellent long-term stability during 5 measurement cycles totally lasting for 29 h. During one measurement cycle, practically no degradation of the catalyst could be detected. This result was compared to some studies using different fuels, catalysts, and operational conditions, as shown in Table 4.3. In most studies, the distinct catalyst deactivation was linked to carbon deposition, which was obviously not a serious problem for the present catalyst during the described measurement cycle. TEM images and EDX measurements of fresh and spent catalyst clearly showed that the crystalline structure and morphology was not changed during operation and no carbon deposition could be detected after usage in a disk-shaped reactor for 40 h. This finding is essential for achieving meaningful investigations of the catalyst and especially for using the catalyst in a practical application.

| Fuel processing | Cracking of pyrolysis bio-oil | $CH_4$ dry reforming | $CH_3OH$ steam reforming | LPG steam reforming | $C_4H_{10}$ partial oxidation |
|---|---|---|---|---|---|
| Catalyst | $Rh/CeZrO_x$ | $Ni/AlZrO_x$ $Ni/AlO_x$ $Ni/ZrO_x$ | $Cu/ZnAlO_x$ | Ceria * Ceria ** $Ni/AlO_x$ | $Rh/CeZrO_x$ |
| Temp. | 700°C | 700°C | 260°C | 900°C | 550°C |
| Time | 12 · 5 min | 6 h | 6 h | 6 h | 6 h |
| Charact. parameter | $X_{H_2}$ | $CH_4$ conversion | $CH_3OH$ conversion | $S_{H_2}$ | $\eta$ $S_{H_2/CO}$ |
| Catalyst deactivation | 12% | 20% 24% 58% | 18% | 5% 17% 36% | <1% |
| Reference | [63] | [64] | [65] | [7] | this study |

Table 4.3: Long-term stability of a disk-shaped reactor for $\phi = 0.8$ and $\dot{V}_{\text{gas,in}} = 20$ sccm compared to different results from literature. *: high specific surface area ($SSA_{\text{BET}}$), **: low specific surface area ($SSA_{\text{BET}}$).

## 4.5.2 Effect of C/O ratio and inlet flow rate

Increasing the C/O ratio had a negative effect on the butane conversion, simply because the amount of $C_4H_{10}$ available to react increased. On the other hand, a higher C/O ratio increased the efficiency of the fuel processing or in other words, increased the selectivity towards $H_2$ and CO, since TOX consuming more $O_2$ than POX was suppressed by the relative lack of $O_2$, as it has been shown by earlier studies, e.g. [66]. These two counteracting effects resulted in maxima of the $H_2$ and CO yields between $\phi = 0.7$ and 0.9: for higher C/O ratios, the conversion of $C_4H_{10}$ was too low and for lower C/O ratios, the production of $H_2$ and CO was not efficient.

Increasing the inlet flow rate $\dot{V}_{\text{gas,in}}$ led to shorter residence time and therefore lower $C_4H_{10}$ conversion, but higher $H_2$ and CO selectivities. The effect of increased selectivities by shorter residence time has also been discussed by other researchers for different hydrocarbon fuels, e.g. [28].

In a previous study [55], it was proven that Rh/ceria/zirconia nanoparticles have a promising potential as catalyst for butane-to-syngas processing, providing higher $C_4H_{10}$ conversion and

| Fuel processing | Catalyst | Temp. | $\eta$ | $S_{H_2}$ | $Y_{H_2}$ | $S_{CO}$ | Ref. |
|---|---|---|---|---|---|---|---|
| $C_4H_{10}$ SR | Pd/ceria | 450°C | 11% | 6% | - | - | [38] |
| $C_4H_{10}$ SR | Pd/ceria | 527°C | 41% | - | - | - | [52] |
| $C_4H_{10}$ SR | Pt/CeGdO$_x$ | 600°C | 63% | - | 39% | - | [53] |
| $C_4H_{10}$ POX | Pt foam | 1041°C | 99% | - | 31% | 79% | [6] |
| LPG POX | Ceria | 700°C | 97% | 42% | - | 15% | [7] |
| $C_4H_{10}$ POX | Rh/CeZrO$_x$ | 550°C | 95% | 80% | 66% | 65% | this study |

Table 4.4: Catalytic performance of DSPB-10 for $\phi = 0.8$ and $\dot{V}_{gas,in} = 20$ sccm compared to different results from literature: butane conversion $\eta$, $H_2$ selectivity $S_{H_2}$, $H_2$ yield $Y_{H_2}$, and CO selectivity $S_{CO}$.

selectivities towards syngas than other catalysts. The catalytic activity of the novel disk-shaped packed bed reactor presented in this study can be compared to results from different studies using $C_4H_{10}$ [6, 38, 52, 53] or $C_3H_8/C_4H_{10}$ mixtures (LPG) [7] as fuel. Only for temperatures of 700°C and higher, other catalysts could partly compete with the results of this study, as shown in Table 4.4. Considering the much lower operating temperature of 550°C, the herein presented Rh/ceria/zirconia DSPB showed excellent feasibility for syngas production from $C_4H_{10}$. This reduction in operating temperature by several hundred °C is essential for using this catalyst in a practical application such as a micro SOFC system running at intermediate temperatures.

### 4.5.3 Comparison of a disk-shaped and a tubular reactor

The comparison of a tubular and a disk-shaped packed bed reactor proved that the disk-shaped design, providing more cross sectional area along the flow direction, resulted in improved catalytical activity and much lower pressure drop in the packed bed. The advantage in catalytic behavior could be explained by the proportionally larger catalytically active area towards the outlet of the reactor, where large parts of the inlet $C_4H_{10}$ and all $O_2$ were already consumed. The increasing cross sectional area influenced the pressure drop significantly, since the flow rate in the reactor increased along the flow direction due to the reactions taking place.

### 4.5.4 Reaction pathways in different reactor regions

Due to the inlet gas composition of $C_4H_{10}$ and dry air, the dominating reactions in the beginning of the reactor were of oxidative nature. POX producing directly $H_2$ and CO was competing in this region with TOX. The well-known indirect syngas production by $H_2O$ consuming reactions such as SR and WGS reducing the $H_2O$ content of the product gas has been suggested as an important reaction path, e.g. by Kunimori et al. [37] and Wang and Gorte [38]. Nevertheless, the results presented in this study show that not only $H_2O$ consuming but as well $CO_2$ consuming reactions followed after TOX of butane and produced syngas. The large consumption of $CO_2$ in Regions 2 and 3, as shown in Fig. 4.6, as well as the simultaneous production of $H_2$ and CO indicated that $CO_2$ or Dry Reforming (DR) was enhanced next to SR. Although DR has been shown to be an efficient way of hydrocarbon reforming when used as an initial reaction, e.g. by Laosiripojana and Assabumrungrat [7], the simultaneous syngas production by POX, SR, and DR is a significant novel finding. The results presented in this study suggest that syngas production by $H_2O$ and $CO_2$ consuming reactions could top syngas produced by POX even when dry air was used as the only oxidizing inlet gas and $H_2O$ and $CO_2$ could only be produced by preceding TOX. Laosiripojana and Assabumrungrat [67] proved that ceria-based catalysts are favorable for DR as well as for SR. In this study, Rh/ceria/zirconia was proved to be an excellent catalyst for oxidative reactions as well as for SR and DR. The simultaneous increase in $H_2$ and CO production could be seen as a sign that WGS was of minor importance in the present reactor.

The undesired production of $CH_4$ mainly at the end of the reactor seemed to be mass transfer limited. By increasing the total inlet flow rate, the relative amount of generated $CH_4$ decreased. The large production of $H_2$ and CO in Region 3 contradicted the possibility of methanation of $H_2$ and CO and rather suggested the generation of $CH_4$ by decomposition of $C_4H_{10}$, as it has been shown for the processing of different hydrocarbons, e.g. for octane by [68].

The disk-shaped packed bed reactor provided a very efficient solution to combine both direct syngas production by POX and indirect pathways via SR and DR. The fast oxidative reactions took place in a small part of the reactor close to the inlet (less than 3 mm radius). The gas flow was radially slowed down despite the increase in gas flow due to the chemical reactions, by the increase in cross sectional area in the flow direction. This allowed for slower reactions such as SR and DR to produce more syngas in the presence of a catalyst that enhances POX as well as the indirect syngas production.

### 4.5.5 Effect of catalytic surface area and catalyst site density

The results presented before showed clearly that the conversion of $C_4H_{10}$ to $H_2$ and CO strongly depended on both the total amount of catalytically active surface sites $n_{Rh,s}$ and the surface site density of Rh $\rho_{Rh,s}$. The $C_4H_{10}$ conversion was highest for DSPB-10 with the highest $\rho_{Rh,s}$ (except for $\phi = 1.2$) and the selectivities towards $H_2$ and CO were highest for DSPB-40 with the highest $n_{Rh,s}$. Although the number of active Rh surface sites of DSPB-80 (0.802 $\mu$mol) was much higher than that of DSPB-10 (0.493 $\mu$mol) and closer to that of DSPB-40 (0.962 $\mu$mol), the catalytic activity of DSPB-80 was significantly lower, especially for higher inlet flow rates. This could be explained by the low Rh site density $\rho_{Rh,s}$ of DSPB-80 compared to the other reactor configurations. A large number of catalytic sites did not lead to satisfactory catalytic performance unless the surface site density was high enough.

This result has an important impact on numerical simulations of processors converting hydrocarbon to syngas: the number of catalytically active surface sites, often quantified by the Catalyst Space Velocity $CSV = \dot{V}_{gas,in}/n_{Rh,s}$, does not characterize the reactor adequately if the surface site density $\rho_{Rh,s}$ is completely neglected. Especially in the case of catalyst site densities remarkably lower than that of a fully covering crystalline layer ($\rho_{Rh,s} = 27.2 \ \mu mol/m^2$) as in the here-presented reactors, $\rho_{Rh,s}$ has to be taken into account in addition to $n_{Rh,s}$.

## 4.6 Conclusion

The present study investigated the production of syngas from butane, using a disk-shaped packed bed reactor containing Rh/ceria/zirconia nanoparticles at 550°C. The disk-shaped reactors achieved high selectivities towards $H_2$ and CO up to 92% and 82%, respectively, and complete $C_4H_{10}$ conversion, depending on the C/O ratio and total inlet flow rate. Besides this very high catalytic activity, the long-term stability of the catalyst was shown to be excellent during the investigated operation period. The disk-shaped packed bed reactor demonstrated significant advantages of catalytic activity and a 6.5 times lower pressure drop compared to a conventional tubular packed bed reactor. This increased catalytic performance was due to a remarkably high contribution of Steam Reforming and Dry Reforming following Total Oxidation next to initial Partial Oxidation. This threefold $H_2$ and CO production pathway on one single catalyst material is a significant novel discovery and led to excellent catalytic results. It was shown that an appropriate characterization of the catalytic performance of the reactors is affected by both the catalytically active surface area and the surface density of catalytically active sites.

The introduced disk-shaped reactor proved to have a high potential for application in micro SOFC systems achieving excellent catalytic performance at a low operating temperature and reasonable pressure drop within a small and compact reactor volume.

# Chapter 5

# Catalytic ceramic foam

Parts of this chapter are published in:
N. Hotz, N. Koc, T. Schwamb, N.C. Schirmer, D. Poulikakos, Catalytic porous ceramic prepared in-situ by sol-gelation for butane-to-syngas in micro-reactors, AICHE JOURNAL 55(7) (2008).

## 5.1 Abstract

In this study, a novel flow-based method is presented to place catalytic nanoparticles into a reactor by sol-gelation of a ceramic foam consisting of Rh/ceria/zirconia nanoparticles, silica sand, ceramic binder, and a gelation agent. This method allows for the placement of a liquid foam precursor containing the catalyst into the final reactor geometry without the need of impregnating or coating of a substrate with the catalytic material. The so-generated ceramic foam shows properties highly appropriate for its use as catalytic material in (micro-)reactors, e.g. reasonable pressure drop due to its porosity, high thermal and catalytic stability, and excellent catalytic behavior.

To investigate the catalytic activity, micro-reactors containing this ceramic foam are employed for the production of hydrogen and carbon monoxide-rich syngas from butane. The effect of operating parameters such as the reactor volume and the inlet flow rate on the hydrocarbon processing are analyzed, showing that a reduced residence time increases the syngas production significantly for sufficiently large reactors.

## 5.2 Introduction

Ceramic foams or sponges are seen as a very promising material for structured catalyst supports due to their porosity, open-cell structure, thermal and chemical stability, and variable thermal conductivity depending on the support material. The typical fabrication procedure of such a catalytic foam is to first generate a porous ceramic as support and then coat this foam in a second step with the catalyst using different coating methods, e.g. shown by [69]. A very common method is the impregnation of the calcined ceramic foam with a liquid precursor containing the catalyst, as presented by [70–72]. An interesting idea is to fabricate the ceramic foam support by sol-gelation, e.g. [71], or to use a sol-gelation method to apply the catalyst on the ceramic foam support, as done by Liguras et al. [73].

The novelty of the herein presented method is to fabricate a ceramic foam containing a catalyst in form of nanoparticles in a direct one-step method by sol-gelation avoiding any separate coating or impregnation step [74]. Reitzmann et al. [70] claimed that a one-step method is usually not possible due to the high sintering temperatures necessary for sufficient mechanical stability of the foam and that therefore ceramic foams typically have to be coated with the catalyst in a second step. However, the present method results in excellent stability of the foams avoiding sintering at high temperature.

The basic idea of this study is to use flow principles for the placement of catalysts in reactors made possible by a sol-gelation method for direct fabrication of a catalytically active ceramic foam from a paste- or gel-like precursor, allowing for the direct precise placement of the catalytic foam in a reactor. This promises several advantages compared to the conventional method of coating or impregnating a rigid ceramic foam with a catalytic material in a second step after producing the inert ceramic foam in a first step. First, the catalyst is added to a liquid carrier and not to an already rigid foam. This is advantageous since the catalyst can be introduced to the foam support material very homogeneously and in a well defined manner by simple mixing and stirring in contrast to the conventional method of impregnating a foam with a liquid catalyst precursor.

Second, this method avoids the existence of loose or dry particles during the production of the foam as well as during its catalytic operation after drying. The usage of loose particles is delicate or even impossible for clean room production techniques and might cause severe health risks. The method presented herein eliminates the problem of polluting the ambient by nanoparticles during production and operation. The dried foam sticks very well together and does not require any kind of filters or ceramic fiber plugs to fix its position. The catalytic nanoparticles cannot erode from the reactor, which is a crucial problem in packed bed reactors consisting of loose particles.

Finally, the foam can be transported using flow principles in the form of a paste-like precursor to its desired position on a substrate or in a reactor. Especially when dealing with a reactor with varying cross section (typically a smaller cross section at the inlet and the outlet of the reactor) in small-scale applications, this is a crucial advantage since a rigid foam cannot be introduced into such a closed reactor. Closing the reactor after introducing the foam is often difficult under clean room conditions. Due to the thermal treatment of the initially wet foam, the dry and rigid foam adheres well to the substrate or reactor wall without any gap or void space in between. This avoids bypassing of gas around the catalytic foam.

To test the catalytic behavior of such a sol-gelation foam containing catalytic nanoparticles, butane-to-syngas processing is performed. This processing has already been undertaken with the same nanoparticles in the form of packed beds with loose particles [55]. The idea is to use such a small-scale butane-to-syngas processor as part of an entire micro SOFC system [19]. When using butane and dry air as inlet gas mixture, Partial Oxidation (POX), written as

$$C_4H_{10} + 2\,O_2 \rightarrow 5\,H_2 + 4\,CO, \qquad (5.1)$$

will be the most desirable reaction path for syngas production from hydrocarbons. Another effective reaction for hydrocarbon processing is Steam Reforming (SR), where butane reacts with water:

$$C_4H_{10} + 4\,H_2O \rightarrow 9\,H_2 + 4\,CO. \qquad (5.2)$$

A second water consuming reaction which might take place in a micro-reactor besides SR is Water Gas Shift (WGS):

$$CO + H_2O \rightarrow CO_2 + H_2. \qquad (5.3)$$

A well performing butane processor should show high selectivity towards POX products instead of Total Oxidation (TOX) products of butane, which is written as

$$C_4H_{10} + 6.5\,O_2 \rightarrow 5\,H_2O + 4\,CO_2. \qquad (5.4)$$

## 5.3 Experiments

### 5.3.1 Preparation of ceramic foam

Nanoparticles made of $Ce_{0.5}Zr_{0.5}O_2$ doped with 2.0 wt% rhodium are prepared in a one-step process by flame spray synthesis as described in previous studies [40, 47, 49, 55]. To a dry mixture containing 24.1 wt% Rh/ceria/zirconia nanoparticles (average diameter: 10 nm), 72.2 wt% silica sand (Riedel-deHaen, average diameter: 200 $\mu$m), 0.9 wt% citric acid salt (triammonium citrate, purum, $\geq$97.0%, Riedel-deHaen) as gelation agent, and 2.8 wt% sodium metasilicate pentahydrate (purum, ≥97.0%, Riedel-deHaen) as ceramic binder a similar mass of distilled water is added. The mixture is mechanically (manually) stirred and finally placed in an ultrasonic bath, both steps for about 5 minutes each. The relatively large silica sand is used as a kind of buffer material to avoid hot spots due to its thermal conductivity, to increase the average pore size, and therefore, to decrease the pressure drop caused by gas flow through the foam. The gelation agent causes a gelation or foaming process which leads to a foam at dried state with a significantly higher porosity than a comparable packed bed of loose particles. The ceramic binder helps all catalytic and $SiO_2$ particles adhere together and to the reactor wall.

By mixing the solid components with a liquid carrier and by mechanical stirring, a gel- or paste-like, highly viscous suspension can be generated. This paste can be easily applied on a substrate or into a cavity by simply forcing it to flow with a pressurized gas, by printing technologies (e.g. ink jet printing), or by any other mechanical or flow application methods. Once the paste is placed on a substrate or in a reactor cavity, the paste is thermally treated by heating it up to 100°C at a low heating rate of 2.5°C/min and kept at this temperature for 2 hours to evaporate all water inside the gel. The remaining rigid ceramic foam does not need any further thermal or chemical treatment.

### 5.3.2 Reactor

This novel method to produce a catalytic foam is demonstrated by a foam reactor in a quartz glass tube (length: 30 cm, inner diameter: 2 mm, outer diameter: 4 mm), where the gel-like foam precursor is pushed to the center of the tube by pressurized air. The tested tubular foam reactors have a reactor volume of 30 (shown in Fig. 5.1), 17, and 12.5 mm³ and accordingly a reactor length of 9.6, 5.4, and 4.0 mm. When pushing the wet gel through the tube, no residues of the foam material remain on the tube wall, as it can be seen in Fig. 5.1.

Figure 5.1: Schematic of the foam reactor in a quartz glass tube

### 5.3.3 Catalyst characterization

The catalytic Rh/ceria/zirconia nanoparticles are characterized in a previous study [55] using the Brunauer-Emmett-Teller (BET) method, Transmission Electron Microscopy (TEM) and X-ray Diffraction (XRD) techniques. The number of catalytically active Rh surface sites is determined by hydrogen chemisorption. The dried foam reactors are characterized by Scanning Electron Microscopy (SEM) techniques and the used nanoparticles are analyzed by TEM and Energy-dispersive X-ray spectrometry (EDX) techniques.

### 5.3.4 Test setup

Butane (PanGas, 3.5, 99.95%) is mixed with synthetic air (79% $N_2$, 21% $O_2$, PanGas, 5.6, purity of both species: 99.9999%), both at 2.5 bar. The flow rates are controlled by Low Delta-P flow meters (Bronckhorst), allowing to operate the reactor slightly above the ambient pressure. The butane/air mixture is fed into the reactor tube placed inside a large tube furnace (MTF 12/38/250, Carbolite). The reactor tube is heated over a length of 30 cm, ensuring practically isothermal conditions inside the foam reactor placed in the middle of the furnace. The product gas leaving the furnace is maintained at around 115°C to avoid condensation of water. The gas composition is analyzed by a gas chromatograph (6890 GC) coupled with a mass spectrometer (5975 MS, Agilent), using a HP-MOLSIV and a HP-PlotQ column (Agilent), respectively. Helium (PanGas, 5.6, 99.9996%) is added as an internal standard for GC calibration. Under typical run conditions, the molar product gas balances of C, H, and O are closed within 5%.

### 5.3.5 Catalytic testing procedure

The reactor is heated from room temperature up to 550°C at a heating rate of 12.5°C/min. Before the reactor reaches the nominal oven temperature of 550°C and the inlet flow of air/butane mixture is started, it is flushed with 20 sccm of air for at least 10 min. Each operation point is kept for at least 20 min before the GC/MS measurement is started to ensure steady state. After the measurements, the tube oven is turned off and the reactor is flushed again with 20 sccm of air for about 1 h while cooling down to 350°C.

The experimental results are analyzed using characteristic values quantifying the catalytic behavior of the packed bed reactor, based on the mole fractions of the outlet gas measured by the GC/MS. The butane conversion $\eta$ of the micro-reactor is determined as the molar ratio between converted butane and inlet butane,

$$\eta = \frac{\dot{n}_{C_4H_{10},\text{in}} - \dot{n}_{C_4H_{10},\text{out}}}{\dot{n}_{C_4H_{10},\text{in}}}. \tag{5.5}$$

The selectivities for hydrogen and carbon monoxide read

$$S_{H_2} = \frac{\dot{n}_{H_2,\text{out}}}{\dot{n}_{H_2,\text{out}} + \dot{n}_{H_2O,\text{out}}} \tag{5.6}$$

and

$$S_{CO} = \frac{\dot{n}_{CO,\text{out}}}{\dot{n}_{CO,\text{out}} + \dot{n}_{CO_2,\text{out}}}. \tag{5.7}$$

To quantify the efficiency of the fuel processor, the exergy of inlet and outlet flows are compared, using the definition of flow availability $a$ from [20]. Since the herein-presented reactor is considered as a component of an entire micro SOFC system, hydrogen and carbon monoxide are assumed to be theoretically fully usable by the SOFC and the exergetic efficiency $\mu_{H_2+CO}$ only considering $H_2$ and CO is calculated by

$$\mu_{H_2+CO} = \frac{a_{H_2,\text{out}} + a_{CO,\text{out}}}{a_{\text{tot,in}}}, \tag{5.8}$$

where the inlet flow availability is practically solely the chemical exergy of the inlet butane. Since an SOFC is probably able to convert a certain amount of $CH_4$ and $C_4H_{10}$ in presence of $H_2O$ by internal reforming, the efficiency of the reformer might consider the exergy output of other species than $H_2$ and CO, which is accounted for by the total exergetic efficiency,

$$\mu_{\text{tot}} = \frac{a_{\text{tot,out}}}{a_{\text{tot,in}}}, \tag{5.9}$$

where the real exergetic efficiency of the reformer is between $\mu_{H_2+CO}$ and $\mu_{\text{tot}}$, depending on the performance of the SOFC itself. The C/O ratio or equivalence ratio $\phi$ based on POX, calculated as

$$\phi = \frac{2 \cdot \dot{n}_{C_4H_{10},\text{in}}}{\dot{n}_{O_2,\text{in}}}, \tag{5.10}$$

is kept constant at 0.8 for all measurements, which has been suggested as an optimal operating point for butane processing in previous studies, e.g. [55]. The total inlet flow rate $\dot{V}_{\text{gas,in}}$ is varied during experiments, resulting in a variation of space time $\tau$, defined as the ratio of reactor volume and total volumetric flow rate at the reactor inlet,

$$\tau = \frac{V_{\text{reactor}}}{\dot{V}_{\text{gas,in}}}, \tag{5.11}$$

and used as an approximation of the residence time. Analogously, a catalyst space time $\tau_{cat}$ can be calculated to quantify the molar amount of catalyst per gas flow rate:

$$\tau_{cat} = \frac{n_{cat}}{\dot{V}_{gas,in}}. \tag{5.12}$$

Since the number of catalyst sites per reactor volume is identical for all tested reactors, the catalyst space time is directly proportional to the space time: $\tau_{cat} \sim \tau$.

## 5.4 Results

### 5.4.1 Structural analysis of ceramic foam

The density of the reactors is determined as $1.25$ g/cm$^3$, whereas the density of the solid material is $3.13$ g/cm$^3$, leading to a volumetric porosity of $60.0\%$. Accordingly, the tested reactors of 30, 17, and 12.5 mm$^3$ contain 9.0, 5.1, and 3.8 mg of catalytic Rh/ceria/zirconia nanoparticles, respectively.

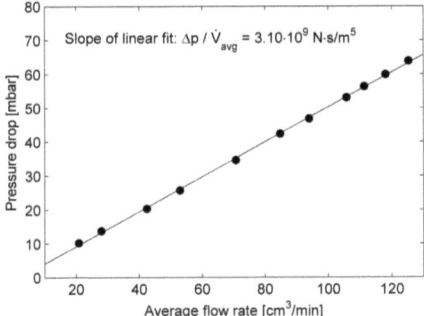

Figure 5.2: Pressure drop through the foam reactor during butane-to-syngas processing as a function of the averaged gas flow rate.

The pressure drop during the operation of the foam reactor is measured continuously and the average pressure drop for each flow rate is shown for one example (30 mm$^3$ reactor volume) in Fig. 5.2. The flow rate $\dot{V}$ is averaged between the known inlet and outlet conditions. Due to the low velocities in the reactor, the pressure drop $\Delta p$ shows good agreement to the linearity proposed by Darcy's Law,

$$\Delta p = \frac{v \cdot \rho}{\kappa} \cdot \frac{L}{A} \cdot \dot{V}, \tag{5.13}$$

and using kinematic viscosities $v$ and densities $\rho$ averaged between inlet and outlet of the reactor, the permeability $\kappa$ amounts to $3.42 \cdot 10^{-11}$ m$^2$, where $A$ is the cross sectional area and $L$ the length of the foam reactor.

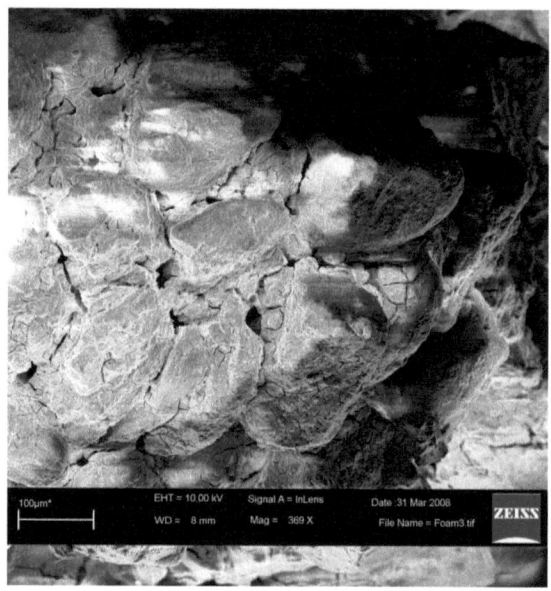

Figure 5.3: SEM image of the foam showing silica sand particles in the order 200 $\mu$m and larger pores between them.

In SEM images, the structure of the silica particles of 200 $\mu$m average diameter can be distinguished very well with pores of some tens of $\mu$m between them (Fig. 5.3). On the surface of these large silica particles, cracks and pores of the order of 1 $\mu$m can be detected (Fig. 5.4). The relatively large $SiO_2$ particles are homogeneously covered with a thin layer of Rh/ceria/zirconia nanoparticles (Fig. 5.5).

Figure 5.4: SEM image showing cracks and pores in the catalytic layer covering the larger silica particles.

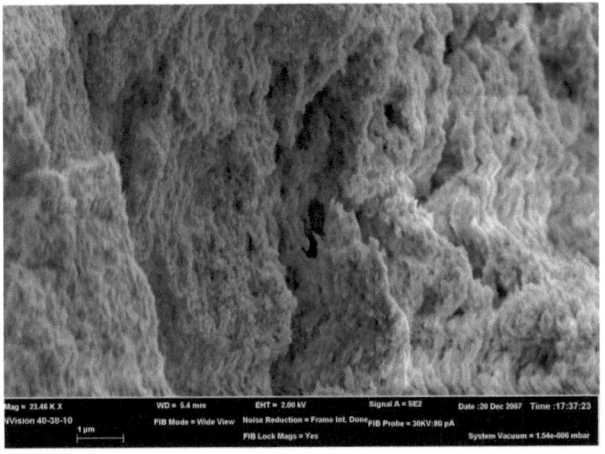

Figure 5.5: SEM image showing the catalytic layer consisting of nanoparticles with a porous surface.

## 5.4.2 Catalytic stability test

Before investigating the effect of operational parameters on the catalytic performance of the foam reactors, the catalytic stability is tested for a foam reactor of 30 mm$^3$ volume. Within the first hour of operation, the catalytic behavior is slightly unstable. After one hour however, the catalytic performance is very stable for several days of measurement. In Table 5.1, results are shown for 3 different inlet flow rates, each measured during at least 5 h. Butane conversion, hydrogen selectivity, and carbon monoxide selectivity show excellent stability during these measurements with a maximum relative error of 0.52%. Foam reactors have been tested for up to 40 h of operation without showing any deactivation of the catalyst.

| Inlet flow rate | Butane conversion (%) | | | H$_2$ selectivity (%) | | | CO selectivity (%) | | |
|---|---|---|---|---|---|---|---|---|---|
| | $\overline{\eta}$ | $\sigma_\eta$ | rel.err. | $\overline{S}_{H_2}$ | $\sigma_{H_2}$ | rel.err. | $\overline{S}_{CO}$ | $\sigma_{CO}$ | rel.err. |
| 20 sccm | 86.5 | 0.45 | 0.52 | 88.1 | 0.27 | 0.30 | 67.2 | 0.29 | 0.43 |
| 30 sccm | 84.2 | 0.20 | 0.24 | 88.5 | 0.11 | 0.13 | 69.6 | 0.26 | 0.37 |
| 40 sccm | 84.0 | 0.34 | 0.40 | 89.3 | 0.13 | 0.15 | 72.7 | 0.16 | 0.22 |

Table 5.1: Average ($\overline{i}$), standard deviation ($\sigma_i$), and relative error (rel.err. = $\sigma_i/\overline{i}$) of butane conversion, hydrogen selectivity, and carbon monoxide selectivity for inlet flow rates of 20, 30, and 40 sccm.

The structure and the surface of the nanoparticles after 40 h of operation are analyzed by taking TEM images clearly showing no change of the chemical composition and the structure of the nanoparticles compared to TEM images of the fresh catalyst (Fig. 5.6). Carbon deposition on the catalyst cannot be detected on any sample by TEM and EDX measurements.

## 5.4.3 Effect of inlet flow rate, reactor volume, and space time

In Fig. 5.7, the measured catalytic performance for three different reactor volumes as a function of the inlet flow rate of butane/air mixture is shown. The butane conversion decreases with increasing flow rate and with lower reactor volumes (Fig. 5.7(a)). For the two smaller reactors, the conversion drops rapidly with increasing flow rates and the measurements are stopped at flow rates between 13 and 19 sccm, since the butane conversion is too low for a practical application in these cases. Nevertheless, the butane conversion of a larger reactor of 30 mm$^3$ is significantly higher and stays between 83% and 84% even for the highest flow rates. The selectivities towards hydrogen and carbon monoxide show similarly an essentially different behavior for the smaller reactors of 17 and 12.5 mm$^3$ compared to the larger reactor (Fig. 5.7(b) and 5.7(c)): The smaller reactors feature lower selectivities and reach maximal selectivities for flow rates between 5 and 10 sccm, whereas no maximum is reached for the 30 mm$^3$ reactor within the tested range. The highest hydrogen selectivities for the smaller reactors amount to about 81% and for the larger reactor, the hydrogen selectivity increases almost linearly from 85% to 89%

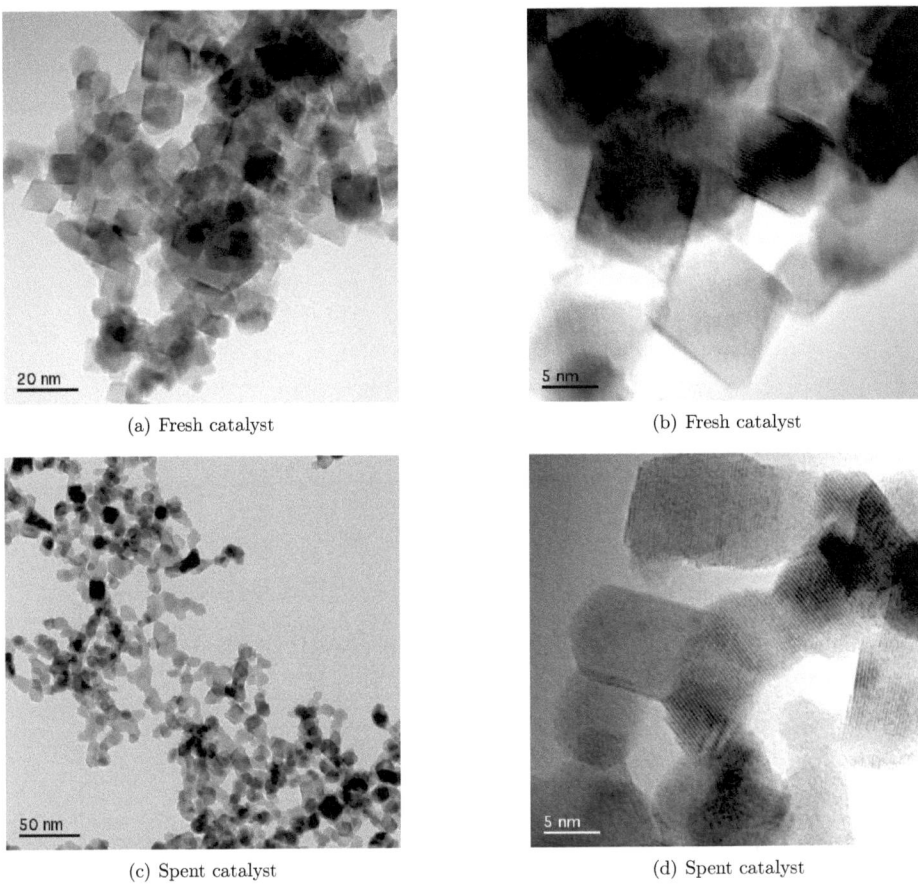

Figure 5.6: TEM images showing the catalyst (a) and (b) before usage and (c) and (d) after 40 h of operation.

for the tested flow rates, which is an excellent result considering the low operating temperature of 550°C. The maximum carbon monoxide selectivity changes for all reactor volumes, reaching 46% for 12.5 mm$^3$ and 50% for 17 mm$^3$ reactor volume. For the largest reactor, almost 73% carbon monoxide selectivity can be achieved.

For the smaller reactors, the maxima in hydrogen and carbon monoxide selectivity between 5 and 10 sccm result in maximal exergetic efficiency $\mu_{H_2+CO}$ of around 38% in this flow range, as it can be seen in Fig. 5.7(d), if only hydrogen and carbon monoxide are accounted for. For the reactor of 30 mm$^3$, the exergetic efficiency $\mu_{H_2+CO}$ increases from 48% to 59% for the highest flow rates. The total exergetic efficiency $\mu_{tot}$ is significantly higher than $\mu_{H_2+CO}$ mainly due to

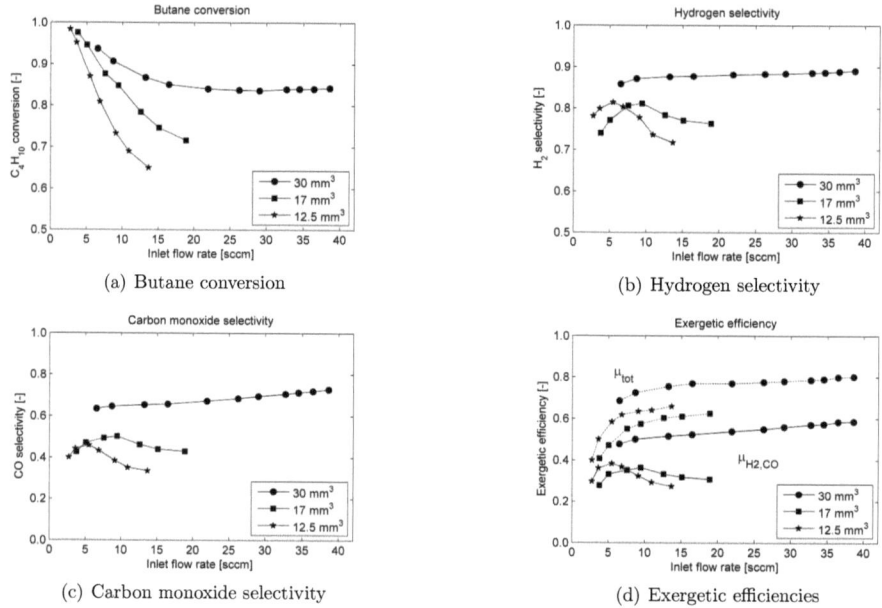

Figure 5.7: (a) Butane conversion, (b) hydrogen selectivity, (c) carbon monoxide selectivity, and (d) exergetic efficiencies $\mu_{H_2+CO}$ and $\mu_{tot}$ of foam reactors with 30, 17, and 12.5 mm$^3$ reactor volume for different flow rates.

unconverted butane.

Next, the same results are plotted as function of space time to better account for the different reactor volume. It can be seen very well that the space time, and in our case therefore also the catalyst space time, are not the only parameters determining the catalytic performance. The results of Fig. 5.8 show that the catalytic characteristics change for different reactor volumes at constant space time. The butane conversion decreases with decreasing space time, however, this decrease is much lower for larger reactors (Fig. 5.8(a)). This results in higher butane conversion for smaller reactors at high space times (above 60 ms) and higher butane conversion for larger reactors at short space times (below 40 ms). The drastic drop in conversion seen for the two smaller reactors does not occur for the reactor of 30 mm$^3$, where the butane conversion stagnates at around 84% for space times down to 17 ms. The selectivities towards hydrogen and carbon monoxide of the smaller reactors reach maxima between 40 and 50 ms (Fig. 5.8(b) and 5.8(c)). For the largest reactor, both selectivities increase monotonically for shorter space times without reaching an optimum within the investigated range of flow rates and space times.

Since the inlet gas flow of butane as fuel increases for shorter space times, the absolute exergy of the product gas usable for the SOFC in form of hydrogen and carbon monoxide grows with

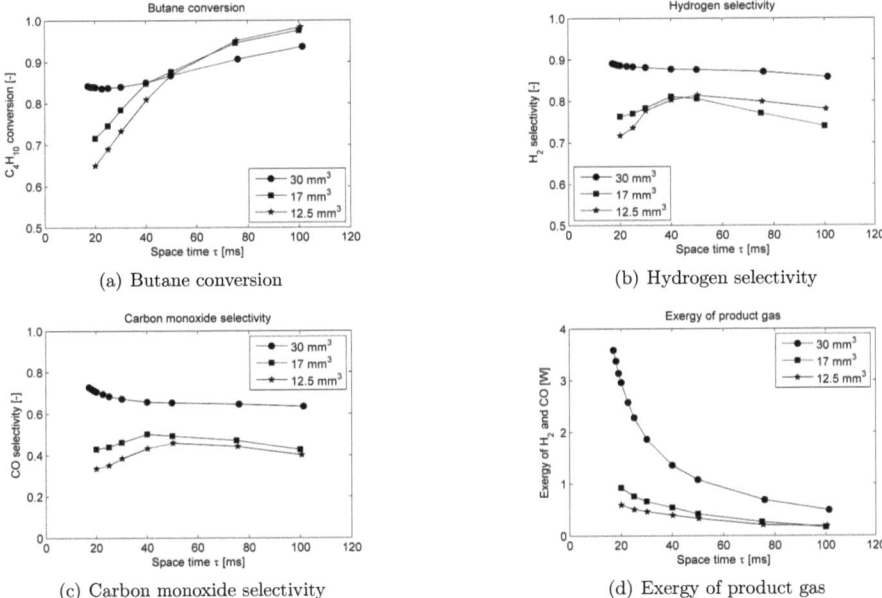

Figure 5.8: (a) Butane conversion, (b) hydrogen selectivity, (c) carbon monoxide selectivity, and (d) exergy of the product gas in form of hydrogen and carbon monoxide for foam reactors with 30, 17, and 12.5 mm$^3$ reactor volume for different space times.

decreasing space time and increasing reactor volume, as it can be seen in Fig. 5.8(d). For the 30 mm$^3$ reactor, hydrogen and carbon monoxide with an exergy or flow availability of 3.58 W can be produced at a space time of 17 ms, indicating a power density of 120 MW/m$^3$. For the smaller reactors, the exergy output in form of hydrogen and carbon monoxide is still 0.93 and 0.59 W, respectively, leading to power densities of 54 and 47 MW/m$^3$. For the largest foam reactor, the exergy output increases dramatically for space times below 40 ms.

## 5.4.4 Reactor temperature

To achieve isothermality, the reactor tube is placed in a large tube furnace operating at a nominal temperature of 550°C. The outer surface area of the reactor tube is relatively large compared to the small foam reactor (see Fig. 5.1) and the low gas flow rates, allowing for considerable heat transfer from the tube to the ambient, and quartz has a fairly high thermal conductivity. The silica sand inside the foam helps to avoid local hot spots inside the reactor and enhances thermal dissipation.

To prove the assumption of a practically isothermal reactor, the temperature is measured by three thermocouples. The first one is positioned 1 mm upstream in front of the foam, the two other thermocouples are placed right at the surface of the foam at the inlet and the outlet. The temperatures are shown for different inlet flow rates of butane/air mixture up to 35 sccm and a reactor volume of 30 mm$^3$ (Fig. 5.9). Without any gas flow (or with only air flowing through the foam reactor), the temperatures amount everywhere to 545.5°C, slightly below the nominal temperature of 550°C. The temperature at the end of the foam stays constant at around 545.5°C for all flow rates, whereas the temperatures at the beginning and 1 mm upstream in front of the foam increase with higher flow rates. However, even for the highest flow rate tested in this study, the temperature at the entrance of the foam is only increased by about 16 K. Considering the nominal temperature of 550°C, this results in a low relative error of 2%, indicating that the assumption of a practically isothermal reactor is correct. Since the most exothermic reactions consume oxygen, the hottest region of the foam can be supposed to be close to the entrance of the foam, exactly where we measure the temperature.

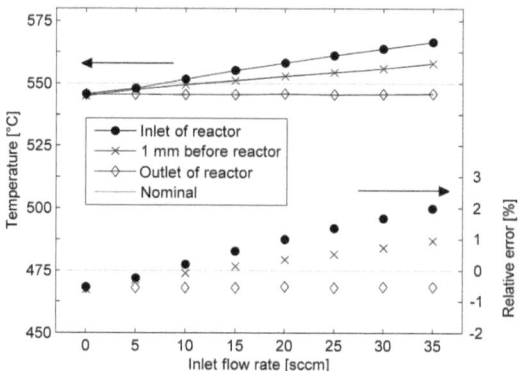

Figure 5.9: Temperatures at three different positions of the foam reactor and the resulting error relative to the nominal reactor temperature as functions of the inlet gas flow rate.

## 5.5 Discussion

### 5.5.1 Structural analysis of ceramic foam

The analysis of the ceramic foam reactors proves that the created foam has very appropriate properties for a catalytic reactor material: an intermediate porosity of 60%, an excellent thermal and chemical stability up to 850°C, and a homogenous distribution of catalytic active material. The porosity is a crucial parameter for an efficient reactor: on the one hand, low porosity and therefore small pore sizes increase the pressure drop of a gas flow through the foam; on the other hand, the void volume has to be limited to provide a large catalytically active surface area. This requirement is fulfilled by the foam due to its multi-scale pores: Larger pores from several tens of microns to the order of 1 μm allow for a high convective mass transport at a reasonably low pressure drop, whereas the nanoporous and completely covering layer of catalytic nanoparticles leads to a large surface-to-volume ratio and a low diffusive mass transport resistance.

Considering the very simple and fast manufacturing procedure, the presented method of producing a catalytic ceramic foam by a direct sol-gelation method in-situ in the final reactor geometry instead of a conventional impregnation of a ceramic supporting foam with a catalyst precursor shows all characteristics necessary for a catalytic reactor material.

### 5.5.2 Catalytic stability test

An essential requirement of a catalytic reactor for significant measurements as well as for potential usage in an industrial application is the catalytic stability. All foam reactors of this study are investigated for at least 40 h of operation without any sign of catalyst deactivation or erosion. Similar to earlier results of the same catalytic nanoparticles [55], the catalyst does not require any pretreatment to reduce the nanoparticles by contact with an $H_2$-rich gas flow, as usually done for such catalysts, e.g. [7, 8, 34, 62]. One advantage of the herein-used catalytic nanoparticles is that no such pretreatment is necessary, since the catalytic activity of the Rh/ceria/zirconia is strong enough from the beginning to convert butane to hydrogen and carbon monoxide. After about one hour of fluctuating catalytic performance, all reactors show very stable operation.

The catalytic stability is a sign that carbon deposition on the catalyst is not a significant problem for the used catalyst for partial oxidation of butane, as it has already been suggested in a previous study [55]. This finding is strongly confirmed by the results of TEM images and EDX measurements proving that not significant carbon deposition takes place. The carbon mass balance for each GC/MS measurement does not show any soot formation in the reactors either.

### 5.5.3 Effect of inlet flow rate, reactor volume, and space time

The results of the catalytic activity show that the structural and chemical properties of the foam supporting the catalytic nanoparticles do not reduce or interfere with its catalytic performance. The catalytic activity of the foam reactors is as excellent as for packed bed reactors containing the same catalyst as loose particles, as shown in [55]. Maximal selectivities towards hydrogen of 89% and carbon monoxide of 74%, respectively, at a simultaneous butane conversion of 84% for the 30 $mm^3$ reactor and the highest inlet flow rate of 40 sccm, referring to a short contact or space time of 17 ms, prove excellent catalytic activity and the ability to efficiently process butane. The production of hydrogen and carbon monoxide within a micro-reactor with a calculated exergy density of 120 MW/m$^3$ is a remarkable result, considering the low reaction temperature of 550°C.

An interesting finding of this study is the coupled effect of inlet flow rate, reactor volume, and space time. The simplistic assumption of constant reactor performance for constant space time and catalyst space time is shown to be incorrect for the tested reactors. The smaller reactors of 17 and 12.5 mm$^3$ reactor volume behave quite similarly, with a certain shift of their maximal selectivities depending on the space time. Nevertheless, their results are well comparable. If the reactor size is further increased to 30 mm$^3$, the reaction characteristics change dramatically. From the results shown here, it can be expected that the selectivities towards hydrogen and carbon monoxide reach maxima for the 30 mm$^3$ reactor at flow rates significantly above 40 sccm inlet flow rate, which is beyond the range tested for this study. From these results it is clear that the reactors are not operated in a mass transfer limited regime, since the space time is not necessarily the limiting factor of the reaction performance. The absolute amount of catalyst seems to be the determining parameter, although the number of catalytically active sites per gas flow is kept constant for constant space times. Above a critical amount of catalyst, the catalytic activity increases to remarkably higher values. The largest reactor of 30 mm$^3$ containing a surface area of 0.81 m$^2$ with 0.44 $\mu$mol of active rhodium surface sites is obviously above this critical value.

This difference in reactivity for different reactor sizes at constant space times can not be explained by a change in temperature of the reactors since all measurements take place at practically isothermal conditions, as shown in subsection 5.4.4.

### 5.5.4 Reactor temperature

An essential objective of this study is to investigate the processing of butane at a relatively low temperature of 550°C without large temperature gradients in the reactor. This is important for an appropriate analysis of the catalytic behavior of the foam and especially critical for a potential application in small electronic devices. The results clearly show that the reactors can be operated at practically isothermal conditions with maximal temperature differences of 16 K, indicating a small relative error of about 2% when compared to the nominal operating temperature of 550°C. This is due to the reactor design allowing for sufficient heat transfer between the reactor and the ambient via thermal conduction through the reactor tubes in axial and radial direction. Although most studies on hydrocarbon processing focus on well insulated or practically adiabatic reactors where necessarily large temperature gradients occur within the reactor and in the gas flow itself, e.g. [28, 75, 76], the configuration chosen for this work reflects the necessity of small temperature differences in practical applications to reduce thermal stress and thermomechanical failure.

## 5.6 Conclusion

In this study, a novel method is introduced to fabricate a ceramic foam containing catalytic nanoparticles by a direct and one-step sol-gelation procedure. With this simple and fast method, a liquid foam precursor already containing the catalyst can be introduced by flow techniques into the final reactor geometry, avoiding the common procedure of impregnating or coating a substrate with the catalytic material. The ceramic foam made by sol-gelation demonstrates excellent properties for a catalytic reactor material, e.g. a low pressure drop due to its porosity of approximately 60% and good thermal and chemical stability.

To prove the high catalytic activity and stability of the foam reactor, the reaction of butane to syngas is investigated on Rh/ceria/zirconia nanoparticles in foam reactors. The maximal hydrogen selectivity amounts to 89% and the carbon monoxide selectivity to 74% at a butane conversion of 84% for the largest tested reactor of 30 mm$^3$ reactor volume and the highest inlet flow rate of 40 sccm. This corresponds to a flow availability in form of hydrogen and carbon monoxide of 3.58 W at an exergetic efficiency $\mu_{H_2+CO}$ of 59%, which is an excellent result for the low operating temperature of 550°C.

The effect of operating parameters such as the reactor volume and the inlet flow rate on the hydrocarbon processing are analyzed. It is shown that the catalytic performance increases significantly for larger reactors, although the space time and catalyst space time are kept constant.

# Chapter 6

# Conclusions

The present work investigates the capability of rhodium doped ceria/zirconia nanoparticles as catalyst for the production of syngas from butane. The main issue of this study is to develop a high catalytic performance at intermediate temperatures of 550°C.

In the range of 500 to 600°C, a packed bed reactor with catalyst nanoparticles of 2.0 wt% Rh loading and $SiO_2$-based plugs achieves nearly complete butane conversion with a hydrogen yield of 77%. This results in an off-gas containing a $H_2$ mole fraction $X_{H_2}$ of 21% and a CO mole fraction $X_{CO}$ of 13%. In spite of its wide use as a sealing material or monolith base, the influence of $Al_2O_3$ in the sealing plugs of the reactor strongly affects the overall performance and suggests a more detailed investigation of non-noble metal catalyzed and homogeneous contributions to butane conversion in such reactors. The first results of the Rh/ceria/zirconia nanoparticles support the further investigation of this catalyst.

In a next step, the production of syngas from butane is investigated by using a disk-shaped packed bed reactor containing Rh/ceria/zirconia nanoparticles at 550°C. The disk-shaped reactors achieve high selectivities towards $H_2$ and CO up to 92% and 82%, respectively, and complete $C_4H_{10}$ conversion, depending on the C/O ratio and total inlet flow rate. Besides this very high catalytic activity, the long-term stability of the catalyst is shown to be excellent during the investigated operation period.

The disk-shaped packed bed reactor demonstrates significant advantages of catalytic activity and a 6.5 times lower pressure drop compared to a conventional tubular packed bed reactor. This increased catalytic performance is due to a remarkably high contribution of Steam Reforming and Dry Reforming following Total Oxidation next to initial Partial Oxidation. This threefold $H_2$ and CO production pathway on one single catalyst material is a significant novel discovery and leads to excellent catalytic results. It is shown that an appropriate characterization of the catalytic performance of the reactors is affected by both the catalytically active surface area and the surface density of catalytically active sites.

The introduced disk-shaped reactor proves to have a high potential for application in micro SOFC systems achieving excellent catalytic performance at a low operating temperature and reasonable pressure drop within a small and compact reactor volume.

Finally, a novel method is introduced to fabricate a ceramic foam containing catalytic nanoparticles by a direct and one-step sol-gelation procedure. With this simple and fast method, a liquid foam precursor already containing the catalyst can be introduced by flow techniques into the final reactor geometry, avoiding the common procedure of impregnating or coating a substrate with the catalytic material. The ceramic foam made by sol-gelation demonstrates excellent properties for a catalytic reactor material, e.g. a low pressure drop due to its porosity of approximately 60% and good thermal and chemical stability.

To prove the high catalytic activity and stability of the foam reactor, the reaction of butane to syngas is investigated on Rh/ceria/zirconia nanoparticles in foam reactors. The maximal hydrogen selectivity amounts to 89% and the carbon monoxide selectivity to 74% at a butane conversion of 84% for the largest tested reactor of 30 mm$^3$ reactor volume and the highest inlet flow rate of 40 sccm. This corresponds to a flow availability in form of hydrogen and carbon monoxide of 3.58 W at an exergetic efficiency $\mu_{H_2+CO}$ of 59%, which is an excellent result for the low operating temperature of 550°C.

The effect of operating parameters such as the reactor volume and the inlet flow rate on the hydrocarbon processing are analyzed. It is shown that the catalytic performance increases significantly for larger reactors, although the space time and catalyst space time are kept constant.

The present dissertation proves that the here-presented catalyst is well suited to provide small and portable butane processing units for applications together with micro fuel cells in different reactor configurations and structures.

# Nomenclature

## Latin letters

| | |
|---|---|
| $A$ | cross sectional area (m²) |
| $a$ | flow availability or exergy (W) |
| $CSV$ | Catalyst Space Velocity (m³ mol⁻¹ s⁻¹) |
| $d_{BET}$ | BET particle diameter (m) |
| $GSV$ | Gas Space Velocity (s⁻¹) |
| $L$ | length (m) |
| $n_{cat}$ | number of catalytically active surface sites (mol) |
| $n_{Rh,s}$ | number of active Rh sites (mol) |
| $\dot{n}$ | molar flow rate (mol s⁻¹) |
| $\Delta p$ | pressure drop (Pa) |
| $S_{H_2/CO}$ | hydrogen/carbon monoxide selectivity |
| $SSA_{BET}$ | BET specific surface area (m² kg⁻¹) |
| $t_{space}$ | space time (s) |
| $V$ | volume (m³) |
| $\dot{V}$ | volumetric flow rate (m³ s⁻¹) |
| $X_i$ | mole fraction of species i |
| $Y_i$ | yields of species i |

## Greek letters

| | |
|---|---|
| $\eta$ | butane conversion |
| $\kappa$ | permeability (m²) |
| $\mu$ | exergetic efficiency |
| $\rho$ | density (kg m⁻³) |
| $\rho_{Rh,s}$ | active Rh site density (mol m⁻²) |
| $\sigma$ | standard deviation |
| $\tau$ | space time (s) |
| $\tau_{cat}$ | catalyst space time (mol s m⁻³) |
| $\upsilon$ | kinematic viscosity (m² s⁻¹) |
| $\phi$ | equivalence or C/O ratio |
| $\psi$ | hydrogen yield |

# List of Figures

1.1 Energy density per mass and per volume for different fuels and Li-ion batteries. Sources: Perry's Chemical Engineers' Handbook [4], International Energy Agency (IEA). . . . . . . . . . . . . . . . . . . . . . . . . . . . . . . . . . . 2

2.1 Schematic of the OneBat micro SOFC system . . . . . . . . . . . . . . . . . 6

3.1 Schematic of the test setup for catalyst performance measurements of a packed bed reactor . . . . . . . . . . . . . . . . . . . . . . . . . . . . . . . . . . . . 13

3.2 Electron micrographs of flame-made $Ce_{0.5}Zr_{0.5}O_2$ (a) and 0.5 wt% $Rh/Ce_{0.5}Zr_{0.5}O_2$ (b) showed regularly shaped nanoparticles with similar morphology. After sintering at 1000°C for 16 h under air, the $Rh/Ce_{0.5}Zr_{0.5}O_2$ (c) displayed marginal particle growth. Sintering necks between adjacent particles stabilized the open, well accessible material. . . . . . . . . . . . . . . . . . . . . . . . . . . . . . 15

3.3 X-ray diffraction patterns of as-prepared (bottom) and sintered samples (top). After preparation, the material consisted of a homogenous ceria/zirconia solid solution. The severe sintering conditions resulted in partial phase segregation. Bottom: reference pattern of $\kappa$-$Ce_{0.5}Zr_{0.5}O_2$ [48]. . . . . . . . . . . . . . . . . 16

3.4 Butane conversion $\eta$ and hydrogen yield $\Psi$ (a), carbon monoxide selectivity $S_{CO}$ and hydrogen selectivity $S_{H_2}$ (b) as functions of reactor temperature for packed bed reactors with a catalyst loading of 0.5 wt% Rh and for both types of plugs, with $SiO_2$ (●) and $Al_2O_3/SiO_2$ (□) fibers, respectively, for an inlet C/O ratio $\phi$ = 0.8. The butane conversion is as well shown for a packed bed without catalyst consisting only of $SiO_2$ sand and both fiber materials. The solid lines through the data points are curve fits. . . . . . . . . . . . . . . . . . . . . . . . . . . 17

3.5 The influence of Rh loading on $Ce_{0.5}Zr_{0.5}O_2$ on butane conversion $\eta$ (a), hydrogen yield $\Psi$ (b), carbon monoxide selectivity $S_{CO}$ (c), and hydrogen selectivity $S_{H_2}$ (d) for an inlet C/O ratio $\phi$ = 0.8, using plugs of $SiO_2$ fibers. The solid lines through the data points are curve fits. . . . . . . . . . . . . . . . . . . . . . . 19

3.6 Butane conversion $\eta$ and hydrogen yield $\Psi$ (a), carbon monoxide selectivity $S_{CO}$ and hydrogen selectivity $S_{H_2}$ (b) as functions of reactor temperature for packed bed reactors with a catalyst loading of 2.0 wt% Rh and SiO$_2$ fiber plugs in an Inconel reactor tube (●) and a quartz reactor tube ($\triangledown$), respectively, for an inlet C/O ratio $\phi = 0.8$. The solid lines through the data points are curve fits. . . . . 20

3.7 Temperatures of four thermocouples in the reactor (1: at the inlet, 2 and 3: in the packed bed, 4: at the outlet) during one measurement cycle. The small plots show the four temperatures averaged over 5 minutes before each measurement point and referenced to the first thermocouple. . . . . . . . . . . . . . . . . . . . . . . 21

4.1 Schematic of the disk-shaped packed bed reactor . . . . . . . . . . . . . . . . . 29

4.2 Transmission electron micrographs of 2.0 wt% Rh/Ce$_{0.5}$Zr$_{0.5}$O$_2$ nanoparticles (a) and (b) before use and (c) and (d) after 40 h of catalytic operation under butane/air flow at 550°C. . . . . . . . . . . . . . . . . . . . . . . . . . . . . . . . 33

4.3 Disk-shaped reactor: Catalytic performance (butane conversion $\eta$, H$_2$ selectivity $S_{H_2}$, and CO selectivity $S_{CO}$) is shown for 7 measurement cycles, lasting 5.8 h each, for a C/O ratio $\phi = 0.8$ and a total inlet flow rate $\dot{V}_{gas,in} = 20$ sccm. . . . . 34

4.4 (a) Disk-shaped reactor: Butane conversion $\eta$, H$_2$ selectivity $S_{H_2}$, and CO selectivity $S_{CO}$ and (b) H$_2$, H$_2$O, CH$_4$, CO, and CO$_2$ yields are shown for C/O ratios $\phi$ between 0.5 and 1.2 and total inlet flow rates $\dot{V}_{gas,in}$ of 10, 15, 20, 25, and 30 sccm. + indicates the equilibrium state. . . . . . . . . . . . . . . . . . . . . . . 35

4.5 Tubular reactor TPB: Butane conversion $\eta$, H$_2$ selectivity $S_{H_2}$, and CO selectivity $S_{CO}$ are shown for C/O ratios $\phi$ between 0.5 and 1.2 and total inlet flow rates $\dot{V}_{gas,in}$ of 10, 15, 20, 25, and 30 sccm. + indicates the equilibrium state. . . . . . 37

4.6 Molar flow rates of H$_2$, CO, C$_4$H$_{10}$, H$_2$O, CO$_2$, and CH$_4$ are shown for C/O ratios $\phi$ between 0.5 and 1.2 and total inlet flow rates $\dot{V}_{gas,in}$ of 10, 20, and 30 sccm after passing the regions of 6, 8, and 10 mm diameter. . . . . . . . . . . . 40

4.7 Disk-shaped reactors DSPB-10, DSPB-40, and DSPB-80: Butane conversion $\eta$, H$_2$ selectivity $S_{H_2}$, and CO selectivity $S_{CO}$ are shown for C/O ratios $\phi$ between 0.5 and 1.2 and a total inlet flow $\dot{V}_{gas,in}$ of 10 sccm, depending on the catalyst loading. + indicates the equilibrium state. . . . . . . . . . . . . . . . . . . . . . 41

4.8 Disk-shaped reactors DSPB-10, DSPB-40, and DSPB-80: Butane conversion $\eta$, H$_2$ selectivity $S_{H_2}$, and CO selectivity $S_{CO}$ are shown for C/O ratios $\phi$ between 0.5 and 1.2 and a total inlet flow $\dot{V}_{gas,in}$ of 20 sccm, depending on the catalyst loading. + indicates the equilibrium state. . . . . . . . . . . . . . . . . . . . . . 42

4.9 Disk-shaped reactors DSPB-10, DSPB-40, and DSPB-80: Butane conversion $\eta$, $H_2$ selectivity $S_{H_2}$, and CO selectivity $S_{CO}$ are shown for C/O ratios $\phi$ between 0.5 and 1.2 and a total inlet flow $\dot{V}_{gas,in}$ of 30 sccm, depending on the catalyst loading. + indicates the equilibrium state. .................................. 42

5.1 Schematic of the foam reactor in a quartz glass tube ................. 50

5.2 Pressure drop through the foam reactor during butane-to-syngas processing as a function of the averaged gas flow rate. ......................... 53

5.3 SEM image of the foam showing silica sand particles in the order 200 $\mu$m and larger pores between them. ................................... 54

5.4 SEM image showing cracks and pores in the catalytic layer covering the larger silica particles. ............................................. 55

5.5 SEM image showing the catalytic layer consisting of nanoparticles with a porous surface. ..................................................... 55

5.6 TEM images showing the catalyst (a) and (b) before usage and (c) and (d) after 40 h of operation. ............................................ 57

5.7 (a) Butane conversion, (b) hydrogen selectivity, (c) carbon monoxide selectivity, and (d) exergetic efficiencies $\mu_{H_2+CO}$ and $\mu_{tot}$ of foam reactors with 30, 17, and 12.5 mm$^3$ reactor volume for different flow rates. .................. 58

5.8 (a) Butane conversion, (b) hydrogen selectivity, (c) carbon monoxide selectivity, and (d) exergy of the product gas in form of hydrogen and carbon monoxide for foam reactors with 30, 17, and 12.5 mm$^3$ reactor volume for different space times. 59

5.9 Temperatures at three different positions of the foam reactor and the resulting error relative to the nominal reactor temperature as functions of the inlet gas flow rate. ..................................................... 60

# List of Tables

3.1 Comparison of performance for packed beds with $Al_2O_3/SiO_2$ and $SiO_2$ fiber plugs at 525°C and 600°C, respectively, using $Rh/Ce_{0.5}Zr_{0.5}O_2$ nanoparticles with 0.5 wt% Rh. . . . . . . . . . . . . . . . . . . . . . . . . . . . . . . . . . . . 22

4.1 Properties of the catalysts used in three tested disk-shaped reactors . . . . . . . . 32

4.2 Maximal $H_2$ and CO yields and pressure drop of the entire test rig for total inlet flow rates $\dot{V}_{\text{gas,in}}$ of 10, 15, 20, 25, and 30 sccm, comparing DSPB-10 and TPB. . 37

4.3 Long-term stability of a disk-shaped reactor for $\phi = 0.8$ and $\dot{V}_{\text{gas,in}} = 20$ sccm compared to different results from literature.*: high specific surface area ($SSA_{\text{BET}}$), **: low specific surface area ($SSA_{\text{BET}}$). . . . . . . . . . . . . . . . . . . . . . . 43

4.4 Catalytic performance of DSPB-10 for $\phi = 0.8$ and $\dot{V}_{\text{gas,in}} = 20$ sccm compared to different results from literature: butane conversion $\eta$, $H_2$ selectivity $S_{H_2}$, $H_2$ yield $Y_{H_2}$, and CO selectivity $S_{\text{CO}}$. . . . . . . . . . . . . . . . . . . . . . . 44

5.1 Average ($\bar{i}$), standard deviation ($\sigma_i$), and relative error (rel.err. $= \sigma_i/\bar{i}$) of butane conversion, hydrogen selectivity, and carbon monoxide selectivity for inlet flow rates of 20, 30, and 40 sccm. . . . . . . . . . . . . . . . . . . . . . . . . . . 56

# Bibliography

[1] C Stone. Fuel cell technologies powering portable electronic devices. *FUEL CELL BULLETIN*, 10:12–15, 2007.

[2] AC Fernandez-Pello. Micropower generation using combustion: Issues and approaches. *PROCEEDINGS OF THE COMBUSTION INSTITUTE*, 29(Part 1):883–899, 2003.

[3] S Ha, Z Dunbar, and RI Masel. Characterization of a high performing passive direct formic acid fuel cell. *JOURNAL OF POWER SOURCES*, 158(1):129–136, JUL 14 2006.

[4] RH Perry and DW Green, editors. *Perry's Chemical Engineers' Handbook*. McGraw-Hill, 7$^{\text{th}}$ edition, 1997.

[5] L Hartmann, K Lucka, and H Kohne. Mixture preparation by cool flames for diesel-reforming technologies. *JOURNAL OF POWER SOURCES*, 118(1-2):286–297, MAY 25 2003.

[6] M Huff, PM Torniainen, and LD Schmidt. Partial oxidation of alkanes over noble-metal coated monoliths. *CATALYSIS TODAY*, 21(1):113–128, AUG 30 1994.

[7] N Laosiripojana and S Assabumrungrat. Catalytic dry reforming of methane over high surface area ceria. *APPLIED CATALYSIS B-ENVIRONMENTAL*, 60(1-2):107–116, SEP 1 2005.

[8] M Huff and LD Schmidt. Production of olefins by oxidative dehydrogenation of propane and butane over monoliths at short-contact times. *JOURNAL OF CATALYSIS*, 149(1):127–141, SEP 1994.

[9] M Huff and LD Schmidt. Oxidative dehydrogenation of isobutane over monoliths at short-contact times. *JOURNAL OF CATALYSIS*, 155(1):82–94, AUG 1995.

[10] AS Bodke, SS Bharadwaj, and LD Schmidt. The effect of ceramic supports on partial oxidation of hydrocarbons over noble metal coated monoliths. *JOURNAL OF CATALYSIS*, 179(1):138–149, OCT 1 1998.

[11] SD Park, JM Vohs, and RJ Gorte. Direct oxidation of hydrocarbons in a solid-oxide fuel cell. *NATURE*, 404(6775):265–267, MAR 16 2000.

[12] ZL Zhan and SA Barnett. An octane-fueled olid oxide fuel cell. *SCIENCE*, 308(5723):844–847, MAY 6 2005.

[13] NP Brandon, S Skinner, and BCH Steele. Recent advances in materials for fuel cells. *ANNUAL REVIEW OF MATERIALS RESEARCH*, 33:183–213, 2003.

[14] UP Muecke, D Beckel, A Bernard, A Bieberle-Huetter, S Graf, A Infortuna, P Muller, JLM Rupp, J Schneider, and LJ Gauckler. Micro solid oxide fuel cells on glass ceramic substrates. *ADVANCED FUNCTIONAL MATERIALS*, 18(20):3158–3168, 2008.

[15] CD Baertsch, KF Jensen, JL Hertz, HL Tuller, ST Vengallatore, SM Spearing, and MA Schmidt. Fabrication and structural characterization of self-supporting electrolyte membranes for a micro solid-oxide fuel cell. *JOURNAL OF MATERIALS RESEARCH*, 19(9):2604–2615, SEP 2004.

[16] ZP Shao, SM Haile, J Ahn, PD Ronney, ZL Zhan, and SA Barnett. A thermally self-sustained micro solid-oxide fuel-cell stack with high power density. *NATURE*, 435(7043):795–798, JUN 9 2005.

[17] H Huang, M Nakamura, P Su, R Fasching, Y Saito, and FB Prinz. High-performance ultrathin solid oxide fuel cells for low-temperature operation. *JOURNAL OF THE ELECTROCHEMICAL SOCIETY*, 154(1):B20–B24, 2007.

[18] A Bieberle-Huetter, D Beckel, UR Muecke, JLM Rupp, A Infortuna, and LJ Gauckler. Micro-solid oxide fuel cells as battery replacement. *MST NEWS*, 4:12–15, 2005.

[19] A Bieberle-Huetter, D Beckel, A Infortuna, UP Muecke, JLM Rupp, LJ Gauckler, S Rey-Mermet, P Muralt, NR Bieri, N Hotz, MJ Stutz, D Poulikakos, P Heeb, P Mueller, A Bernard, R Gmuer, and T Hocker. A micro-solid oxide fuel cell system as battery replacement. *JOURNAL OF POWER SOURCES*, 177(1):123–130, FEB 15 2008.

[20] N Hotz, SM Senn, and D Poulikakos. Exergy analysis of a solid oxide fuel cell micropowerplant. *JOURNAL OF POWER SOURCES*, 158(1):333–347, JUL 14 2006.

[21] GW Coffey, JS Hardy, LR Pederson, PC Rieke, and EC Thomsen. Oxygen reduction activity of lanthanum strontium nickel ferrite. *ELECTROCHEMICAL AND SOLID STATE LETTERS*, 6(6):A121–A124, JUN 2003.

[22] B Zhu. Functional ceria-salt-composite materials for advanced ITSOFC applications. *JOURNAL OF POWER SOURCES*, 114(1):1–9, FEB 25 2003.

[23] B Zhu, XT Yang, J Xu, ZG Zhu, SJ Ji, MT Sun, and JC Sun. Innovative low temperature SOFCs and advanced materials. *JOURNAL OF POWER SOURCES*, 118(1-2):47–53, MAY 25 2003.

[24] WJ Stark, K Wegner, SE Pratsinis, and A Baiker. Flame aerosol synthesis of vanadia-titania nanoparticles: Structural and catalytic properties in the selective catalytic reduction of NO by $NH_3$. *JOURNAL OF CATALYSIS*, 197(1):182–191, JAN 1 2001.

[25] A Trovelli. *Catalysis by Ceria and Related Materials*. Imperial College Press, 2002.

[26] B Silberova, HJ Venvik, JC Walmsley, and A Holmen. Small-scale hydrogen production from propane. *CATALYSIS TODAY*, 100(3-4):457–462, FEB 28 2005.

[27] A Mitri, D Neumann, TF Liu, and G Veser. Reverse-flow reactor operation and catalyst deactivation during high-temperature catalytic partial oxidation. *CHEMICAL ENGINEERING SCIENCE*, 59(22-23):5527–5534, NOV-DEC 2004.

[28] DA Hickman and LD Schmidt. Production of syngas by direct catalytic-oxidation of methane. *SCIENCE*, 259(5093):343–346, JAN 15 1993.

[29] KT Nguyen and HH Kung. Generation of gaseous radicals by a V-Mg-O catalyst during oxidative dehydrogenation of propane. *JOURNAL OF CATALYSIS*, 122(2):415–428, APR 1990.

[30] R Burch and EM Crabb. Homogeneous and heterogeneous contributions to the oxidative dehydrogenation of propane on oxide catalysts. *APPLIED CATALYSIS A-GENERAL*, 100(1):111–130, JUL 1 1993.

[31] M Xu and JH Lunsford. Oxidative dehydrogenation of propane. *REACTION KINETICS AND CATALYSIS LETTERS*, 57(1):3–11, JAN 1996.

[32] AA Lemonidou and AE Stambouli. Catalytic and non-catalytic oxidative dehydrogenation of n-butane. *APPLIED CATALYSIS A-GENERAL*, 171(2):325–332, JUL 13 1998.

[33] KD Campbell, E Morales, and JH Lunsford. Gas-phase coupling of methyl radicals during the catalytic partial oxidation of methane. *JOURNAL OF THE AMERICAN CHEMICAL SOCIETY*, 109(25):7900–7901, DEC 9 1987.

[34] B Silberova, HJ Venvik, and A Holmen. Production of hydrogen by short contact time partial oxidation and oxidative steam reforming of propane. *CATALYSIS TODAY*, 99(1-2):69–76, JAN 15 2005.

[35] I Aartun, B Silberova, H Venvik, P Pfeifer, O Gorke, K Schubert, and A Holmen. Hydrogen production from propane in Rh-impregnated metallic microchannel reactors and alumina foams. *CATALYSIS TODAY*, 105(3-4):469–478, AUG 15 2005.

[36] S Marengo, P Comotti, and G Galli. New insight into the role of gas phase reactions in the partial oxidation of butane. *CATALYSIS TODAY*, 81(2):205–213, JUN 15 2003.

[37] K Kunimori, T Iwade, H Uetsuka, S Ito, and T Watanabe. Infrared chemiluminescence study of CO produced by partial oxidation of butane on platinum. *CATALYSIS LETTERS*, 18(3):253–259, 1993.

[38] X Wang and RJ Gorte. A study of steam reforming of hydrocarbon fuels on Pd/ceria. *APPLIED CATALYSIS A-GENERAL*, 224(1-2):209–218, JAN 25 2002.

[39] L Madler, HK Kammler, R Mueller, and SE Pratsinis. Controlled synthesis of nanostructured particles by flame spray pyrolysis. *JOURNAL OF AEROSOL SCIENCE*, 33(2):369–389, FEB 2002.

[40] WJ Stark, L Madler, M Maciejewski, SE Pratsinis, and A Baiker. Flame synthesis of nanocrystalline ceria-zirconia: effect of carrier liquid. *CHEMICAL COMMUNICATIONS*, (5):588–589, 2003.

[41] MG Atwell and JY Hebert. Rhodium determination by atomic absorption spectrometry using nitrous ixide-acetylene flame. *APPLIED SPECTROSCOPY*, 23(5):480–482, 1969.

[42] MJ Stutz and D Poulikakos. Effects of microreactor wall heat conduction on the reforming process of methane. *CHEMICAL ENGINEERING SCIENCE*, 60(24):6983–6997, DEC 2005.

[43] AK Chaniotis and D Poulikakos. Modeling and optimization of catalytic partial oxidation methane reforming for fuel cells. *JOURNAL OF POWER SOURCES*, 142(1-2):184–193, MAR 24 2005.

[44] RE Hayes and ST Kolaczkowski. *Introduction to Catalytic Combustion*. Taylor and Francis, 1997.

[45] R Strobel, F Krumeich, WJ Stark, SE Pratsinis, and A Baiker. Flame spray synthesis of $Pd/Al_2O_3$ catalysts and their behavior in enantioselective hydrogenation. *JOURNAL OF CATALYSIS*, 222(2):307–314, MAR 10 2004.

[46] R Strobel, WJ Stark, L Madler, SE Pratsinis, and A Baiker. Flame-made platinum/alumina: structural properties and catalytic behaviour in enantioselective hydrogenation. *JOURNAL OF CATALYSIS*, 213(2):296–304, JAN 25 2003.

[47] WJ Stark, JD Grunwaldt, M Maciejewski, SE Pratsinis, and A Baiker. Flame-made Pt/ceria/zirconia for low-temperature oxygen exchange. *CHEMISTRY OF MATERIALS*, 17(13):3352–3358, JUN 28 2005.

[48] H Kishimoto, T Omata, S Otsuka-Yao-Matsuo, K Ueda, H Hosono, and H Kawazoe. Crystal structure of metastable $\kappa$-$CeZrO_4$ phase possessing an ordered arrangement of Ce and Zr ions. *JOURNAL OF ALLOYS AND COMPOUNDS*, 312(1-2):94–103, NOV 16 2000.

[49] WJ Stark, M Maciejewski, L Madler, SE Pratsinis, and A Baiker. Flame-made nanocrystalline ceria/zirconia: structural properties and dynamic oxygen exchange capacity. *JOURNAL OF CATALYSIS*, 220(1):35–43, NOV 15 2003.

[50] EA Kummerle and G Heger. The structures of C-$Ce_2O_{3+\delta}$, $Ce_7O_{12}$, and $Ce_{11}O_{20}$. *JOURNAL OF SOLID STATE CHEMISTRY*, 147(2):485–500, NOV 1 1999.

[51] BCH Steele. Materials for IT-SOFC stacks 35 years R&D: the inevitability of gradualness? *SOLID STATE IONICS*, 134(1-2):3–20, OCT 2000.

[52] X Wang and RJ Gorte. Steam reforming of n-butane on Pd/ceria. *CATALYSIS LETTERS*, 73(1):15–19, 2001.

[53] CK Acharya, AM Lane, and TR Krause. Kinetic study of the steam reforming of isobutane using a Pt-$CeO_2$-$Gd_2O_3$ catalyst. *CATALYSIS LETTERS*, 106(1-2):41–48, JAN 2006.

[54] S Hilaire, S Sharma, RJ Gorte, JM Vohs, and HW Jen. Effect of $SO_2$ on the oxygen storage capacity of ceria-based catalysts. *CATALYSIS LETTERS*, 70(3-4):131–135, 2000.

[55] N Hotz, MJ Stutz, S Loher, WJ Stark, and D Poulikakos. Syngas production from butane using a flame-made Rh/$Ce_{0.5}Zr_{0.5}O_2$ catalyst. *APPLIED CATALYSIS B-ENVIRONMENTAL*, 73(3-4):336–344, MAY 11 2007.

[56] D Neumann and G Veser. Catalytic partial oxidation of methane in a high-temperature reverse-flow reactor. *AICHE JOURNAL*, 51(1):210–223, JAN 2005.

[57] MJ Stutz, N Hotz, and D Poulikakos. Optimization of methane reforming in a microreactor - effects of catalyst loading and geometry. *CHEMICAL ENGINEERING SCIENCE*, 61(12):4027–4040, JUN 2006.

[58] GJ Panuccio, KA Williams, and LD Schmidt. Contributions of heterogeneous and homogeneous chemistry in the catalytic partial oxidation of octane isomers and mixtures on rhodium coated foams. *CHEMICAL ENGINEERING SCIENCE*, 61(13):4207–4219, JUL 2006.

[59] KA Williams and LD Schmidt. Catalytic autoignition of higher alkane partial oxidation on Rh-coated foams. *APPLIED CATALYSIS A-GENERAL*, 299:30–45, JAN 17 2006.

[60] T Montini, AM Condo, N Hickey, FC Lovey, L De Rogatis, P Fornasiero, and M Graziani. Embedded Rh(1wt.%)@ $Al_2O_3$: Effects of high temperature and prolonged aging under methane partial oxidation conditions. *APPLIED CATALYSIS B-ENVIRONMENTAL*, 73(1-2):84–97, APR 24 2007.

[61] AV Pattekar and MV Kothare. A radial microfluidic fuel processor. *JOURNAL OF POWER SOURCES*, 147(1-2):116–127, SEP 9 2005.

[62] A Schneider, J Mantzaras, and P Jansohn. Experimental and numerical investigation of the catalytic partial oxidation of $CH_4/O_2$ mixtures diluted with $H_2O$ and $CO_2$ in a short contact time reactor. *CHEMICAL ENGINEERING SCIENCE*, 61(14):4634–4649, JUL 2006.

[63] EE Iojoiu, ME Domine, T Davidian, N Guilhaume, and C Mirodatos. Hydrogen production by sequential cracking of biomass-derived pyrolysis oil over noble metal catalysts supported on ceria-zirconia. *APPLIED CATALYSIS A-GENERAL*, 323:147–161, APR 30 2007.

[64] F Pompeo, NN Nichio, OA Ferretti, and D Resasco. Study of Ni catalysts on different supports to obtain synthesis gas. *INTERNATIONAL JOURNAL OF HYDROGEN ENERGY*, 30(13-14):1399–1405, OCT-NOV 2005.

[65] A Kundu, JH Jang, HR Lee, SH Kim, JH Gil, CR Jung, and YS Oh. MEMS-based micro-fuel processor for application in a cell phone. *JOURNAL OF POWER SOURCES*, 162(1):572–578, NOV 8 2006.

[66] G Veser, M Ziauddin, and LD Schmidt. Ignition in alkane oxidation on noble-metal catalysts. *CATALYSIS TODAY*, 47(1-4):219–228, JAN 1 1999.

[67] N Laosiripojana and S Assabumrungrat. Hydrogen production from steam and autothermal reforming of LPG over high surface area ceria. *JOURNAL OF POWER SOURCES*, 158(2, Sp. Iss. SI):1348–1357, AUG 25 2006.

[68] AD Qi, SD Wang, GZ Fu, and DY Wu. Autothermal reforming of n-octane on Ru-based catalysts. *APPLIED CATALYSIS A-GENERAL*, 293:71–82, SEP 28 2005.

[69] B Schimmoeller, H Schulz, SE Pratsinis, A Bareiss, A Reitzmann, and B Kraushaar-Czarnetzki. Ceramic foams directly-coated with flame-made $V_2O_5/TiO_2$ for synthesis of phthalic anhydride. *JOURNAL OF CATALYSIS*, 243(1):82–92, OCT 1 2006.

[70] A Reitzmann, FC Patcas, and B Kraushaar-Czarnetzki. Ceramic sponges - Application potential of monolithic network structures as catalytic packages. *CHEMIE INGENIEUR TECHNIK*, 78(7):885–898, JUL 2006.

[71] F Snijkers, S Mullens, A Buekenhoudt, W Vandermeulen, and J Luyten. Ceramic foams coated with zeolite crystals. *FUNCTIONALLY GRADED MATERIALS VIII*, 492-493:299–303, 2005.

[72] MV Twigg and JT Richardson. Fundamentals and applications of structured ceramic foam catalysts. *INDUSTRIAL & ENGINEERING CHEMISTRY RESEARCH*, 46(12):4166–4177, JUN 6 2007.

[73] DK Liguras, K Goundani, and XE Verykios. Production of hydrogen for fuel cells by catalytic partial oxidation of ethanol over structured Ni catalysts. *JOURNAL OF POWER SOURCES*, 130(1-2):30–37, MAY 3 2004.

[74] N Hotz, D Poulikakos, A Studart, A Bieberle-Huetter, and LJ Gauckler. Porous ceramic catalysts and methods for their production and use. Patent EP 08 012273, 2008.

[75] R Horn, KA Williams, NJ Degenstein, A Bitsch-Larsen, D Dalle Nogare, SA Tupy, and LD Schmidt. Methane catalytic partial oxidation on autothermal Rh and Pt foam catalysts: Oxidation and reforming zones, transport effects, and approach to thermodynamic equilibrium. *JOURNAL OF CATALYSIS*, 249(2):380–393, JUL 25 2007.

[76] G Veser and LD Schmidt. Ignition and extinction in the catalytic oxidation of hydrocarbons over platinum. *AICHE JOURNAL*, 42(4):1077–1087, APR 1996.

Die VDM Verlagsservicegesellschaft sucht für wissenschaftliche Verlage abgeschlossene und herausragende

# Dissertationen, Habilitationen, Diplomarbeiten, Master Theses, Magisterarbeiten usw.

## für die kostenlose Publikation als Fachbuch.

Sie verfügen über eine Arbeit, die hohen inhaltlichen und formalen Ansprüchen genügt, und haben Interesse an einer honorarvergüteten Publikation?

Dann senden Sie bitte erste Informationen über sich und Ihre Arbeit per Email an *info@vdm-vsg.de*.

**Sie erhalten kurzfristig unser Feedback!**

VDM Verlagsservicegesellschaft mbH
Dudweiler Landstr. 99
D - 66123 Saarbrücken
**www.vdm-vsg.de**

Telefon +49 681 3720 174
Fax     +49 681 3720 1749

Die VDM Verlagsservicegesellschaft mbH vertritt

Printed by Books on Demand GmbH, Norderstedt / Germany